P9-BZZ-450

THE COMPLETE CRYSTAL GUIDEBOOK

a practical path to self development, empowerment and healing

Uma Silbey
U-Read Publications,
San Francisco, CA

© 1986 Uma Silbey

Published by
U-read Publications
655 DuBois St.Suite E
San Rafael, CA 94901

Cover, author, and product photography by
Robert Reiter

Illustrations by Sumi deVeuve

Original Sculpture on cover by
Uma Silbey

Cover Design by
Rainbow Canyon

All photos and product designs ©1986 Uma Silbey

Second Printing, March 1987

ISBN 0-938925-00-8

87 88   10 9 8 7 6 5 4 3

Contents

THE COMPLETE CRYSTAL GUIDEBOOK

This book is dedicated to my husband, Ramana Das, and my friend, Eileen Kaufman, whose help and encouragement enabled it to be written. I also want to thank all teachers and lineages who have guided me on my path and thus have indirectly contributed to this work.

To be (crystal) conscious:
Pay attention
And be impeccably honest with yourself

1

Basic Information

INTRODUCTION

THERE ARE MANY reasons to want to work with quartz crystals. Do you want to do healings? Do you want to empower your meditations, your affirmations, and/or your thoughts? You can energize the body and balance its energies. You can develop many psychic abilities, including clairvoyance, clairaudience and traveling on astral and mental planes. You can use the crystals to change many unwanted circumstances in your life and create new ones. Are these of any interest to you? You can meet guides and beings from different dimensions and uncover ESP abilities. Is this your aim?

Most people beginning crystal work have one or more of the above goals in mind when they start working, and all of these will be addressed in this book. However, a curious thing starts to happen as you work with crystals. You start becoming aware of an energy or force or a "potential" higher than yourself. You start becoming aware of and can begin to interact with something very powerful and wonderful. Some call it Spirit, or a Higher Order, or God. Whatever name you call it, it is universal and transcends our limited self. We find that we begin to experience unlimited

1

potential flowing through us as our prior self-imposed boundaries, limitations and ideas of who and what we are begin to crumble. Unlimited energy, vision, love, creativity, contentment and wisdom begin to flow through us as we open to this potential. We begin to rest in the contentment of what some call the True Self, a feeling of who and what we and the universe are that defies pinpointing and definition. In fact, the more we try to describe what we are feeling and what we are coming in contact with, the more obscure it becomes.

Does this sound fantastic, the wishful thinking of an over-active imagination? Perhaps it does to you, if you have had no experience of this. This book will lay out a path, providing detailed exercises and practices with your crystals which will open you to this experience as well as allow you to do the more traditional work with your crystals. You will be empowered to tap into the wisdom, endless knowledge and peace which flows within you, waiting to be discovered and utilized. In the process of doing the crystal work described within these pages, you will see that not only does this development proceed automatically, but the abilities to use the crystals in any way that you can envision also automatically become available to you.

Much of the crystal work that is currently being taught is on the level of "tricks" that can be done with your crystal ,without taking into account any information of a more wholistic nature that is needed to become more effective in your work. This superficial approach is extremely limited -- you would be limited in the work that you could do, and after a while you would become dissatisfied.

What things do you need to know? Because you are working with subtle energy or vibration when working with the quartz crystal, you need to learn about energy systems as a whole, both inside the body and in the environment which surrounds you. For example, Kundalini energy is often awakened when you start working with crystals. You will need to know how to channel this in yourself, to utilize the unleashed energies and powers and learn how to create this in others. There are chakra systems or energy matrixes in the body that become activated as you work, and these can be activated in those you work with. Because the work you will be doing is on the subtle planes, you'll need to know about the astral and mental planes, and how to work with them. These levels are involved whether you are conscious of them or not. The more conscious you become, the more effective you will be. Because you begin drawing tremendous forces of energy through your body as you work, you must learn how to build up and maintain the strength of the nervous system, and how to maintain the strength and health of your physical body.

In short, you will need to expand your consciousness beyond your physical body, beyond the environment, and into other realms. This book will not only provide intellectual information, but also will provide you with concrete methods to experience and become sensitive to what is being spoken of. You will gain knowledge from your own experience, for you do not truly know anything until you experience it.

Crystal work concerns transcending our limited self and limited ego. In the process, we become master crystal workers, not just good talkers. Quartz crystal work is a combination of the mind, the will, the crystal, and the guiding, empowering spirit. To work effectively you must be in a state of surrender, becoming a channel through which the creative force or spirit can do its work. You are merely the vehicle. You become like a hollow tube through which this spirit can flow without impediments of doubt, ego, excess intellectuality, fear, physical weakness, pride or cluttered mind. To become this channel, you need to learn to become centered, to learn what this state is. You need to develop a calm, focused mind.

Exercises will be provided that allow these qualities to develop within you. Exercises will also be given to help you develop sensitivity to the subtle energies and the vibrations of the crystal, the body, the environment. Common obstacles will be addressed with some information about how to overcome them. This work takes some effort on your part and some faith that it is worth doing. Honesty is necessary. Most people are seeking a "way out," not a "way in." The most common reason for not working is forgetting that initial contact with the force larger than yourself, the forgetting of the deep inner yearning to connect with it. Most people are looking for quick, spectacular results or easy panaceas. Though you will experience results right away, it is important to keep on working -- more is yet to come. You can expect deep, long lasting, permanent changes both within yourself and in your abilities with the crystals. It's necessary to be willing to accept a certain amount of discipline and responsibility.

Both obvious and subtle changes will happen as you continue work with the exercises and practice with your crystals. Your body may go through certain changes, making it more sensitive in certain ways. You may experience a separation of the physical and psychic selves and the ability to use one or the other. You may develop an ability to separate from your mind and emotions so that they serve you but do not control you. Psychic energy will become deeply integrated and used in your everyday life. What before seemed to be miraculous will become commonplace. You will see that all miracles operate according to common laws and you will understand their workings. An unlimited flow of creative energy will open up for you that you can direct in any way. You will find opening up in you a deep love that is not emotionalism, rather a deep compassion and understanding of others. You will find opening up within you a strong desire to serve and you will find out how to achieve that. As you progress you will become fearless and content. You will be able to share your happiness with others.

Be willing to go beyond barriers that might become impediments to your progress. What are some of these barriers? Doubt, impatience, excess imagination or wishful imagination, emotionalism mistaken for "mystical experience," rigid thinking and concepts, intellectualism, inconsistency in your practice of the exercises, the need to be right, negativity and depression, discouragement, the need for recognition, and escapism. Though you might face these and other tests, know one thing: You will never be tested beyond your strength, although it may seem so at the time. Keep going, one step at a time. Everything in your lifestyle may change. Your friends may change. Your job may change. Be willing to let go -- better things are in store for you. Don't give up: you are being guided. Have courage and rely on the truth.

There are many systems of crystal work -- probably as many systems as there are people working with crystals. All of them are probably valid. The point is to try them and check your results in the physical universe. Be honest, and see what actually works *for you.* There is no right and wrong method, only what works. With that in mind, no one system is presented in this book, but several systems, and the methods by which you learn to develop a reliance on your inner guidance so you will not have to rely on any one system. You can be spontaneous and do what each situation and each moment demands. Give the methods that you do try an appropriate length of time to test them. For some people, certain techniques will work right away, while for others these techniques will take a couple of weeks, or a month, or six months or more. Give them time to work. In this book, optimal trial periods will be suggested. Use your intuition, or the inner voice to see what is right for you, the method to use in any given situation. If you do not hear this inner voice at first, methods will be suggested that develop your "hearing" and show you what to listen for and how to be quiet enough to hear. *Trust yourself.*

Finally, remember that the crystal is only a tool, a powerful one, but still a tool. Like any tool, you can do the same thing without it, but it helps to make things easier. The crystal is not a god; it helps you to do a certain job. After some time of working with the crystals, you may find that you can do without them just as effectively. Don't be afraid to let go. Put your crystals aside, admire them for their beauty and honor them for the work they helped you do -- and move on. The crystals are not Gods: you are. You are the creator in your own universe.

Follow your own path

That doesn't mean that we don't listen and learn from others' experiences . . .

Uma modeling Isis sterling silver quartz crystal headband (see chapter on Crystal Tools).

TERMINOLOGY AND EXERCISES

Terms such as energy and vibration that are used in this book are meant to explain a particular experience. They are not necessarily used as they would be in a classic scientific sense. However, the scientists and metaphysicians (including crystal workers) are using different words to explain the identical phenomenon.

If you are confused by the number of different exercises included in this book, wondering how many days to do them or in what order, here is a guide for you. As a general rule of thumb, the order of appearance in the book is the sequence in which the exercises are to be done. Do them one at a time until you begin to experience their full effect. This will develop in you the ability to distinguish between the different energy channels which flow through your body. The exercises to develop the chakras should be done in the order in which they are presented. While you are doing these, you can do the kundalini exercises at the same time. Once you have done the exercises in the book you can repeat them at any time you like and in any order. Use them when you think you need them. Start working with your crystals from the beginning and notice how your effectiveness improves as you do each exercise to develop and sensitize yourself.

PHYSICAL DESCRIPTION

Natural quartz crystals, often referred to in ancient tradition as the "veins of the earth," frozen water or frozen light, are formed naturally from the elements silicon and water through a lengthy process involving heat and pressure. They are buried in the earth, usually where the rock is sandstone in nature, or sometimes in streambeds where they have washed down from higher ground after they have been dislodged. They are often found near gold. Varieties of quartz crystal, sometimes called rock crystal, are found all over the world. Currently, the largest numbers of extremely fine crystals are being mined in Arkansas; Herkimer, New York; Mexico and Brazil.

Their natural formation features six sides or faces with a point on one end, or sometimes on two ends. A crystal with a point on one end is referred to as single-terminated. A crystal with a point on both ends is referred to as double-terminated. They are found in clusters of crystals attached to each other in all directions, or as single crystals which have generally broken off the clusters.

Quartz crystal stones are both clear and colored. Each color has its own rate of vibration particular to it and different attributed characteristics and powers. The most widely used crystal is the clear quartz. Also included in the family of quartz crystals are amethyst which are purple in color, blue quartz, rose quartz, citrine quartz, which vary from pale yellow to fiery orange to light brown, green quartz, rutilated quartz, which have in them fine gold or copper colored fibers, and quartz with black, blue or green tourmaline rods inside. Each has its own particular usefulness which will be explained throughout this book.

Energy, in the form of vibration, is projected from each crystal to form a field around it. This is often referred to as the power of the crystal. Each projected field varies in dimension with each crystal. Generally, a small crystal of approximately one-half inch in size will project a field of around three feet. You can see, then, that even the small crystals can be quite powerful. Larger crystals often, but not necessarily, project a field greater in dimension than the small crystals. However, the size of the crystal is not always the determining factor in the size of the projection. The clarity and brilliance of a crystal often are more important in determining the power of it than the size. Generally, a larger crystal can store more energy in it and can handle more energy passing through it. As will be discussed later, there are formations that you can place the crystals in to increase the projected energy field immensely.

Each quartz crystal contains a line of direction along which energy flows when it is transmitted through the crystal. The energy flow in the crystal runs up from the bottom where it enters, up through the crystal and then out through the point. If the quartz crystal has a point on each end, the energy comes and goes in both directions, as in a battery. As it transforms energy, it expands and contracts slightly at differing rates depending on the rate of vibration of the influence to which it is exposed. (This oscillation is what makes the crystal so essential in radio and television broadcasting.)

There are many methods to increase the power of the crystal or to charge it with energy. To charge a crystal can be seen as enlivening the crystal. Put your crystals in the sunlight. Running water will charge up a crystal, so hold them in the ocean waves, or under waterfalls, or rest them in the beds of running streams. Even cold running water from your tap can charge them slightly. Rushing wind can charge a crystal. Basically, anything that makes you feel more lively, or more full of energy will also energize the crystal. You can charge the crystal with certain types of influences that will not only increase its power, but will allow the stored energy to be used later in work that you might want to do. For example, bury your crystal for a length of time in the earth. That will charge it with the strength, grounding and nurturing of the earth energy. Take them to the beach and bury them partially in the sand so that they receive the earth energy, the energy of the ocean wind, and of the sun above. Let your crystals stay outside overnight to be charged with the feminine moon energy, the stars and the soft darkness of the night. See what energies you would like to use in your work and charge up your crystals accordingly. It is usually a good idea to charge up each crystal with only one type of energy and to keep them wrapped until you would like to use them.

If you want to really know crystals, get to know rock . . .
Find a rock cliff and rub your body against it . . .
Meditate on rock and in rock.

CHOOSING YOUR QUARTZ CRYSTAL AND STORING IT PROPERLY

Now that you know something about the physical properties of quartz crystal, how do you proceed to choose one for yourself? As mentioned earler, consider the size, clarity and the brilliance of the crystal. Also look for veils which please you. These are something like gossamer webs or slightly opaque wisps of clouds inside the crystal. Internal fractures, sometimes with prismatic color effects can form miniature landscapes, or doorways which seem to draw you inside the crystal. These are called inclusions. Look for any formations caused by tourmaline rods, or filaments of gold or copper minerals which have meaning for you. Rainbows can energize the crystal with their color and etheric nature. Some crystals seem more dense, some more etheric, seeming to pull you skyward. Look for phantom crystals that are filled with any number of pyramids if you want to use your crystal for sending messages or like an Egyptian influence. (These crystals are rare!) How does the crystal feel to you? Does it seem to radiate warmth or refreshing coolness? Or do you prefer a perfectly clear crystal, regarding anything else as distraction to you? These are all things that add value and meaning to your crystal. Most systems of crystal lore teach that the tip of the crystal should be whole, not chipped, cracked or broken off. It is said that this detracts from or interrupts the energy flow. Often this is true. However, sometimes a crystal is very powerful even though the tip is chipped.

Sometimes the crystals have been cut and polished into particular sizes and shapes that project energy fields or energy flows particular to those shapes. Examples of this are crystal balls whose round shape sets up a corresponding circular field of energy that seems to easily draw you into its center. Other examples are the crystal cross, and the crystal pyramid. Some crystals are polished on one or more faces, but some faces are left rough, creating interesting gazing crystals with the convoluted landscapes inside. Notice, though, whether the lapidary work on the crystal was done consciously, leaving the energy flow intact. Sometimes the energy flow is cut across or cut through, leaving a severely weakened crystal. The lapidary work should leave the integrity of the stone intact, enhancing its power, not diminishing it. Do you want a crystal to carry around with you to handle? If so, then find one that feels good in your hand. Some are shaped to fit the hand specifically for this purpose. Some are deliberately shaped to be used as massage tools.

How are you going to use your crystal? People that intend to do astral work or dream work usually prefer a double-terminated or a herkimer diamond crystal, a particularly brilliant, double-terminated, multi-faceted crystal only found in the area around Herkimer, New York. (Astral work and dream work will be covered later in this book.) If you want to direct energy in one direction with your crystal, choose a wand crystal, or a single-terminated crystal. Crystal clusters are good for energizing a room or the environment around you. Finally, feel the energy of the crystal to see how strong it is.

Ultimately, the deciding factor is this. Which crystal just seems to draw you to it? What one seems almost irresistible? Choose the one that *intuitively* feels right to you. Most of the time it is the one you looked at first!

After you have chosen your crystal, what is the best way to store it? When not in use you should keep your crystals wrapped in a natural fiber. Most prefer cotton or silk or leather. Be conscious of the color that you wrap it in because the color will influence your crystal. What color feels the best to you when you hold the crystal? This is often the color you will want to use. You may want to store your crystal wrapped or unwrapped on an altar or in some sacred or special place that you have set up. That influence of purity and light will be in your crystal. It is protected against intrusion. You may also want to include something with the crystal that has special meaning or power for you. This will influence your crystal with its energy. Generally, a crystal that you have set aside for a specific use should not be touched by others, just as you would not expose it to any other influences that might interfere with its special function.

You might want to leave some crystals exposed for everyone to share in their beauty and unique radiance.

HOW TO WORK WITH QUARTZ CRYSTALS

Modern physicists have affirmed that physical form consists, in essence, not of matter but of energy -- and that the nature of physical material is intrinsically dynamic, in process. In other words, everything that exists is an external manifestation of an energy form, a rate of vibration. Furthermore, everything exists in dynamic cause and effect relationship with everything else. Nothing exists in a vacuum. Therefore, a change in the rate or form of vibration of a particular form in one location creates a corresponding change in the vibrations of other forms in other related locations, which affect other forms, which affect still more forms, etc. This simple cause and effect mechanism is analogous to the example of a stone dropped into a still pond which causes ripples of water to spread out in ever increasing concentric circles around the original impulse.

When we work with quartz crystals we work with the same principle. We create changes or manipulate in some way the vibrations on a subtle, non-physical level to eventually affect the related vibrations on a physical level. We work subtly to manifest physically. We can use quartz crystals to do this because they have extremely high and exact rates of vibration that can be precisely manipulated due to the crystal's tendency to resonate in harmony with any vibration with which it is brought into contact. Knowing that the crystal tends to harmonize its vibrational rate with another, how do we use that property to have the crystal amplify, store, transform, transmute, and focus vibration to create the changes that are associated with the crystal work? Quartz crystal will automatically harmonize and re-create the vibration of any object with which it is placed in direct physical proximity and/or can be directed to do this by the use of our conscious intention. To direct the crystal's activity with our intention, we use our natural knowing to create a particular set of vibrations within ourselves. We then interact with the crystal so as to have it harmonize and resonate with that newly created vibrational set in us. We can then direct that crystal resonance with the use of our focused intention or will to interact with any other vibrational fields of our choosing without the normal limitations imposed by time and space. If there is no resistance encountered in the chosen vibrational field it will be caused to vibrate in harmony with the crystal's vibrations which are stronger, being highly charged with our intention.

In other words, we set up in the crystal a certain current of energy. The wave that carries or transmits that current is intention. The more focused the intention, the more empowered the wave to travel further and effect more changes. Thus, with the use of the crystal we have created a change in a vibrational field which manifests physically in the way that we intended. (Further elaborations of this procedure will appear in the following chapters.)

The crystal will do some things automatically. Just by being in the crystal's proximity, the body and/or environment will feel energized due to its automatic tendency to raise lower vibratory rates up to its own high level. Also, it generates negative ions to create a feeling of a refreshing, harmonious, uplifting atmosphere around it. Any particular vibratory rate with which the crystal has been influenced will soon be reflected in its surroundings. Most of the time, however, in crystal work, the intention of the user is extremely important.

Because all manifestation of being is essentially vibration, as shamans, priests, mystics, and healers have known for ages, quartz crystals can be used to modify thoughts, emotions, our bodies and other physical forms. Negative emotions can be transformed to positive. States of disharmony can be changed to harmony. Our bodies can be energized or healed. Thoughts can be amplified, increasing the power of affirmation, concentration, meditation, intention, and visualization. In place of stress we can generate calm. The uses and benefits of quartz crystals extend as far as the limits of our vision.

You are your own
laboratory . . .
and your own
scientist

CENTERING AND GROUNDING YOURSELF

Before you begin any crystal work you first need to center and ground yourself. What does this mean?

To be centered refers to that state of being in which you are just yourself. Rather than judging yourself to be this or that, what or who, there is just a feeling of being here now. (This feeling of being here now is not a thought but is an experience apart from thought.) When you are centered, it is a feeling of being collected into your center rather than being scattered. (Thus, the term "centered.") This center feels like it is collected around your heart, your navel or sometimes in-between the two. However, in reality, your center has no particular location. It feels like a state of calm receptivity. To the degree that you are centered, your intuitive voice, your will, your higher energy centers and consciousness become available to you. You can only focus and concentrate when you are centered. You have more energy when you are centered. These prior attributes are necessary to do effective crystal work. Throughout the book are many references to being centered. Also included are exercises to do which at first you need to center yourself. A good practice is to automatically center yourself before each crystal method or other exercise whether it is suggested or not. Though practices to clear your mind, develop concentration, work with emotions and strengthen your will also help you to center yourself, the following is a specific technique to use:

EXERCISES TO CENTER YOURSELF

Sit with the spine very straight. Sit in any way that is comfortable -- cross-legged, on your knees, or in a straight-backed chair. Close your eyes. Focus your attention in the center of your chest. Next, begin to breathe with long, deep breaths through your nose. Fill the lungs, hold your breath for a second or two and then let it out, emptying your lungs completely. When you do this, you fill your body with life force and adjust the body rhythms. Use long, deep breaths. Continue this breathing until you feel centered: it generally takes from three to ten minutes.

Another centering method uses sound, which works very quickly. There are many ways to work with sound. Here is one you can use: take some long deep breaths and close your eyes. Ring a bell or gong with a clear piercing, sustained sound. Relax and concentrate on the sound as you ring it continuously. Allow these sounds to carry you into a calm, centered state.

The next step after centering yourself when preparing to do any crystal work is to ground yourself. Grounding creates a secure attachment and connection with the earth. It permits the flow of energy from the earth to move through the soles of your feet and up through your body. This can then be joined with energy flowing from the sky, through the top of your head and down through your body. The coupling of energy from the "heavens" and the "earth" creates the proper balance to do crystal work. The subtle information that you are able to learn and experience as this happens can be manifested and utilized with the crystal. This cannot happen if you are not grounded. If you are not grounded, you feel "spaced out" or nervous and/or hyperactive, and are often less effective in your daily life.

There are many methods to ground yourself. Wearing or having with you smoky quartz crystals will help. Iron pyrite is extremely grounding. Any earth colored stone such as agate or jasper will be grounding. Crystals and/or grounding stones worn as anklets will have a strong grounding effect. Another way to ground yourself is to open the energy meridian points in the middle of the soles of your feet. Imagine roots growing from the bottoms of your feet into the earth as you walk barefoot on the ground. Try the following technique:

GROUNDING YOURSELF

1. Sit still with a straight spine and center yourself. Close your eyes.

2. Imagine a gold cord of light leaving from the bottom of your spine and traveling down into the earth. If you are inside, envision the cord of light going through the floor or floors, then into the earth.

3. Use your breath if you like. With each exhale send increasing amounts of energy in the form of the golden cord down further towards the center of the earth.

4. You may feel heavier or as if your body has expanded. You might feel tingles along the bottom of your spine.

5. If you feel any tension or stiffness, imagine your out breath releasing these blockages. Then slowly rotate your neck and gently flex any other parts of your body that seem tense.

6. Continue this exercise until you feel grounded or for three minutes.

CLEARING AND PROGRAMMING YOUR CRYSTAL

Quartz crystals store vibration which can come from sources as varied as sound, light, touch, emotion, thought, or the physical environment around them. This vibration can, in turn, affect those who come into contact with the crystal. (This process has been explained in more detail earlier.) So, when you first receive your crystal, before you begin to work with it, before and immediately after you use it for healing work, or when any undesirable influence has entered it, as well as whenever it looks dull or seems to lack vitality, you will need to remove the energy imprinted and stored in the crystal. This removal of stored vibration is called *clearing the crystal.*

Many effective methods can be used for clearing crystals. Here are a few methods that you can try. Try them and use the method that works best for you.

Smudging Method

This is a native American method that is effective to clear yourself, others, and the room that you are in, as well as your crystals and other stones.

First, put some sage, cedar, or sweetgrass in a bowl that is heat-resistant. If you like, you can use an abalone or other shell container, or an incense burner. Light a flame to it, fanning or blowing it until you create some fire and lots of smoke. Then move your crystal(s) through the smoke, or fan or blow the smoke over the crystal, your intention being to clear the crystal. Continue until the crystal looks or seems to be more clear. When selecting the sage or cedar to burn, find cedar trees and sage bushes where they grow wild and pick small sprigs. Dry them before you try to burn them. Individual leaves can be used, but they are more difficult to burn than the sprigs. Of course, be sure that you don't select branches that will burst into an overly large flame that might cause damage. Ground sage from your spice store can only be used with a charcoal block and is difficult to work with. Use the leaves from the cedar tree rather than cedar chips. Sweetgrass is a particular type of wild grass that grows in many areas of the country. It is not the grass that grows in your front lawn.

If you cannot find sage, cedar or sweetgrass, you can use sandlewood incense. If you cannot find sandlewood incense use another incense that most appeals to you. If you are clearing yourself or others, fan or blow the smoke entirely over the body, head to foot. Fan the smoke throughout the room to clear it.

Breath Method

This method works best with single crystals rather than clusters. If you are clearing many crystals at once it is also more time consuming than smudging. First, hold the crystal in your left hand with the tip pointing up. Hold the crystal with your left thumb on the bottom and your left first finger on the tip. Hold it about six inches outward from the center of your chest. Next, place your right thumb on any face of the crystal. As you do that, place your right first finger on the opposite face from the one under your thumb. (See illustration.) Focus on the crystal, intending for it to become clear. As you do that, inhale through your nose and exhale forcefully through your mouth in the direction of the crystal. It is as if your breath carries your intention. After that, place your right thumb on the side next to the one which it covered before. Again, place your right first finger on the opposite face. Inhale and forcefully exhale into the crystal, intending it to be clear. Finally, switch your right thumb and first finger to the remaining two faces. Again, inhale and exhale into the crystal. The crystal is now clear.

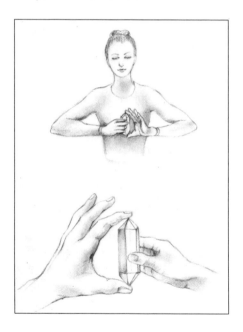

Salt And Salt Water Method

Place your crystal(s) in sea salt for a period of one to seven days. Either place your crystals partially in the salt or bury them entirely. (You can purchase sea salt from your local health food store.) Take them out of the salt when they physically look or intuitively seem more clear. Change your salt at least once a month. If you choose to use the salt water method, you can put your stones in a glass container of water and sea salt. Then, store this container in sunlight for one to seven days. Use about three tablespoons of salt to one cup of water. Use enough water to completely cover the stones. Both of these salt methods rely on the smaller crystals of salt to draw out that which is stored in the large crystal. When you are through clearing the crystals throw the salt water out. (Do not drink it or throw it into your favorite plant.)

Other Methods

You can also use a tape demagnetizer to clear your crystals. Run it up and down the length of your stones. It is possible to use *visualization* to clear your crystals. However, this method is not as reliably effective as the above methods. If your concentration or intention loses its strength or focus, the stones will not be completely cleared. If you do choose to use a visualization method, you intend to clear the crystal(s) as you visualize golden light from the rays of the sun streaming through the tip and out the bottom. This golden light chases all grey negativity out from the bottom of the crystal into the earth where it is transmuted. This leaves your crystal shining and clear.

Now that your crystal is clear you can begin working with it. The remainder of this book describes the process and exact methods by which you can make use of your will, mind, and emotion to affect the vibrations in the crystal(s) to create appropriate changes in subtle, then physical, bodies. The changes in vibration that you caused in a crystal can be used immediately to create changes or can be stored in the crystal to be used later. Or certain vibrations can be stored in the crystal to create effects over as long a length of time necessary until the crystal is cleared. The process of consciously creating a vibration or set of vibrations in a crystal and storing them in the crystal for later and/or continuing effect is known as *programming the crystal*. A crystal can be programmed with thought, emotion, sound, color, touch, or any other influence by which you can normally change the vibration in a crystal.

PROGRAMMING YOUR CRYSTAL

1. Clear the crystal that is to be programmed.
2. Hold the crystal in both of your hands while gazing into it. (If the crystal cannot be picked up, just lay both hands upon it.)
3. Center yourself and clear your mind. Concentrate on that with which you intend to program the crystal.
4. As you retain your concentration, inhale and forcefully exhale through your mouth. It is as if you are blowing your intention into the crystal.
5. Continue this process until you feel satisfied that you have completely filled the crystal with your intention.
6. This vibration is now stored in the crystal until you clear it out of the crystal. Some methods feel that a program is "locked" into a crystal more effectively when steps one through five are done in front of a flame. Then when you are through programming the crystal, you pass it right to left through the flame. This seals in the intended programming.
7. When the programming is done, clear yourself, and the environment around you. When you want the program to be out of the crystal, just clear the crystal. (This is even true if the program is "locked in.")

This describes a method for programming your crystal for immediate or later use. Throughout the rest of the book more information is given that will enable you to completely understand how a crystal can be affected and used, as well as how it can affect our bodies, thoughts, emotions and environment.

Truth is beyond right and wrong . . .
it just is itself
Make truth your guide
This is what is meant by Discrimination

DEVELOPING YOUR SENSITIVITY TO PHYSICALLY FEEL AND INTUITIVELY SENSE VIBRATION

In working with crystals you can intuit the vibrational fields surrounding the crystal, the vibrational patterns that are being affected with its use, as well as the vibrational fields associated with whatever you wish to charge your crystal with. However, particularly with healing work, it is helpful to be able to physically feel the vibrations you are working with. To develop the sensitivity of your hands to be able to feel this energy, try the following exercise:

BREATH AND CRYSTAL SENSITIZATION EXERCISE

Rub the palms of your hands together fiercely for thirty seconds or a minute, generating lots of heat. Then sensitize your hands by opening them and lightly blowing on them to create a tingling sensation. Next, holding your crystal in one hand, lightly touch the point of the crystal to the center of your other hand's palm. Then draw the crystal up and move it around in a circular motion until you can feel the energy, like a tingling or coolness. See how far away you can draw your crystal and still feel the vibration in your palm. Then experiment by running the crystal over your body at a distance of about six inches, experiencing, as you do, the extraordinary power of the energy produced.

As you can see, to feel the vibration produced by the crystal you need to become quite still. In fact, the ability to feel the energy is not only a matter of the sensitivity of the hands, but also a matter of the strength of your mental focus. To achieve the mental focus that you need, you must become still both in mind and body. Allow the mind to become like a still body of water, letting no attachment to runaway thoughts disturb its surface. If you find that your thoughts are wandering, drop your attention from them and gently return your attention to the crystal and your hand. Some people are able to immediately feel the energy produced. Others take more time. If you don't experience success the first time, do it again, or keep on trying several times every day. You will eventually feel the vibration.

This next exercise develops not only your sensitivity to the crystal vibration, but also builds the sensitivity in your hands to feel the subtle vibrational field of physical objects -- yours and others' bodies included. To experience this, try the following exercise:

EXERCISE TO FEEL THE SUBTLE VIBRATIONAL FIELD

First, sensitize your hands using the breath and crystal sensitization exercise as before. Then, while holding the crystal in your hand, take the same hand and run it over a physical object that you have selected. Run your hand at a distance of about six inches away from the object. The crystal will continue to amplify the sensitivity of your hands as you do this. When you can feel the vibration of the energy field around the object, try pulling the hand slowly further and further away until you can no longer feel the vibrational field. Your mind should be still and your attention completely focused as you do this. How far does the energy field around the object seem to extend? The vibrancy and life force of an object is shown by the amount of extension of this vibrational field. (See the information about the astral and mental auras later in the book.)

You will find that as you work to develop the sensitivity of your hands and the concentration of your mind, your intuitive abilities seem to be increasingly called upon. As you use your intellectual mind less, you begin to rely on other ways of knowing that seem to naturally open up for you. As you physically feel more subtle energies you also begin to develop a subtle sensing. It is a feeling of knowing something without knowing how you knew. This sensing does not seem based on any intellectual reasoning, but later can be intellectually confirmed to be correct. Learn to rely on this sensing. That does not mean that you ignore the information coming from the physical universe. In fact, you use the physical universe to test the legitimacy of what you have sensed intuitively. This, in turn, reinforces and builds the accuracy of your intuitive sense. The important thing to do in terms of developing this subtle sense is first to notice it, then to trust it. Trust what you sense and don't be afraid to act on it. Have courage. You will find that you will be able to use the intuitive, subtle sense as well as to physically feel the subtle vibration to do your crystal work. This makes it even more effective.

After you have spent some time developing the ability to feel the vibrational field of physical objects with your crystal, try doing it without the crystal. Do the breath and crystal sensitization exercise, and then run your hand over the vibrational field of an object without using your crystal. Can you still feel it? If not, try again after repeating the crystal and breath sensitization exercise. If you still do not feel it, as in the other sensitization exercises, try several times every day until you do. Eventually you will have success.

As with any activity, constant practice brings improvement. Disuse brings some loss of the ability, although usually the new-found sensitivity of the hands does not entirely disappear. If you loose the ability to physically feel vibration because of lack of practice, it will not take as long to develop as it did the first time. Just practice again.

Once you have developed the ability with one hand, you will want to develop it in the other hand because you usually use both hands in your crystal work. You can develop one hand's sensitivity first, then the other, or you can develop both hands concurrently.

You may want to do the following exercise along with the other sensitization practices. This exercise develops not only the sensitivity of the hands but also stimulates the thyroid and parathyroid secretions to send energy into the upper centers of the head. This will stimulate the pituitary gland to open up and increase the intuitive capacities. The heart center will be energized, and the mind will clear to allow more concentration. While doing this exercise, it is important to hold the position exactly because it creates certain pressures (as you will become aware of), which stimulate reactions that cause the alteration of thought patterns. The sounds you will make also activate certain centers that create the effects you will experience. The exercise should be done for six minutes. If you cannot do it this long at first, start with one minute. Then do three minutes. After you can do three minutes, work up to six minutes. This exercise is not easy. It takes some effort on your part. Use your will power to keep going. As you will experience, the effort is worth it. To completely experience the benefits of this exercise, you should do it for thirty days.

EXERCISE TO DEVELOP INTUITION AND SENSITIVITY OF THE HANDS

Sit in a place where you will not be interrupted. Sit with a straight spine with the head facing forward and the chin level. If you sit on a chair, your feet should be flat on the ground and uncrossed. Extend the arms out to the sides parallel to the ground. Palms are flat and face up. Now, concentrate on the top center of the head and at the same time be aware of the energy in the palms. (You may at first be unable to be aware of both the head and hands simultaneously. (Picture A) As you continue the exercise with this view in mind, this will start happening for you.) If your thoughts wander, when you notice this, bring them gently back to the exercise. After a while, your thoughts will more easily remain focused on the exercise. If you like, you can envision a line of energy from the palms to the top center of the head and from palm to palm. This forms a triangle. Begin with the head facing forward. Next turn it to the left four times saying the sound WHAHO (wah-ho) on each turn. (Picture B) (Say the sound out loud.) After each head motion, the head returns to the forward-facing position. (Picture C) Then turn the head to the right four times using the sound GURU (goo-roo). Again, return the head to face forward each time. Continue in a regular rhythm for six minutes. (The sound will be rhythmic and continuous with no break.) A single repetition lasts about seven seconds. After the six minutes, inhale and exhale deeply and relax the arms. Continue to sit still for a few minutes to integrate the changes.

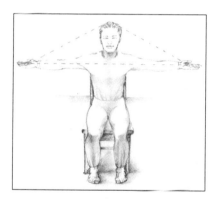

A

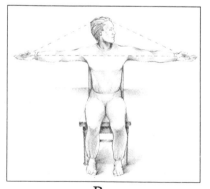

B   C

TRAINING THE MIND

In most people, the mind drifts endlessly from thought to thought, half formed and half followed. These thoughts are triggered by sights we see, sounds we hear, sensual impressions, memories, and by those thoughts that vibrate around us endlessly. Unguarded and untrained, the mind resonates with whatever other thought vibrations happen to be in the environment around us or directed to us. As the uncontrolled mind resonates with these various vibratory states around us, we are influenced by them. Thus, if people around us are angry, we become angry. If they are happy, we are happy, etc. If the sky is overcast, we may feel depressed. Busy city vibrations make us feel tense and terse. We may dwell in our memories rather than in the present, those memories creating certain emotional states that create the corresponding thoughts.

All too often we become the slaves of our minds, rather than the masters. When we need to focus the mind, the "muscles" are not there. We cannot hold a thought or image, or still the mind. The more we try to still the mind, the busier it becomes. We are unable to concentrate, our thoughts endlessly chasing after each other until we are finally distracted by some other stronger thought. This describes the state of the average mind, the undeveloped mind.

What is the alternative? What do we mean when we speak of the developed mind? The developed mind can be likened to a still body of water. Thoughts, like waves, cross its surface, but the water itself is undisturbed. Thoughts come and go, but they do not ruffle the mind. The mind rests in a state of peaceful awareness. When the mind is then called upon to direct its attention in some way, it does so with strength and steadiness. No distractions deflect its attention which is held by our will. When the developed mind is focused, it is aware of nothing else but what it is focused on. It is held steady. When the need for focus is over, the mind then returns to its prior state of balanced equilibrium. Only when the mind is quiet with its endless discriminations and judgments stilled can we become aware of the continuing stream of wisdom which flows through us to guide us in our work.

What is the need for a clear and steady mind in crystal working? We have seen that the crystal has the capability to influence vibration to manifest changed conditions. This can be used to our advantage. We create a particular vibration of thought that the crystal resonates in harmony with and then send that thought vibration to interact with another set of vibrations thus changing them as we intended. With this technique, we can make changes in the mental, astral, etheric and physical planes. To do this effectively, we need to be sure that the visualization or thought or intention is held unwaveringly for the amount of time that is needed to accomplish the results. The thoughts or visions need to be focused with strength so that the crystal receives their impressions clearly, not being confused with other competing thoughts or visions. Only one set of vibrations is transmitted. If the mind is cluttered and uncontrolled there is no way that this can be done. There is no control over the other thoughts that impinge upon the intended envisioning. Furthermore, the images or thoughts need to be held steadily while the will is applied to direct them through the crystal to be transmitted in the way intended. The more focused our minds are able to be, the more in our control, the more effective our work will be.

The following is a technique to use to develop and train your mind. Try it and then test for the difference in your crystal work. It is recommended to try the technique for at least thirty days to see results. After doing this technique to train your mind, try the third practice called Exercise to Transmit Peacefulness. In this exercise you will use your mind and intention to direct vibration with a crystal to create change.

FOLLOWING YOUR BREATH EXERCISE

Sit in a place and situation where you will not be disturbed. Be relaxed but have your spine straight. You may lay down flat on your back; however, this has a tendency to cause you to sleep. Close your eyes and bring your attention to your breathing. Continue to breathe naturally and notice where your breath seems to tickle either the tip of the nose, the front of the nostrils, or the upper lip. Instead of a tickle, you may feel this as a slight coolness. If you do not notice where the air passes on its way in or out of your body, continue to focus on the tip of the nose or the front, top of the lip until you feel the air's passage. As you continue to sit upright in a relaxed manner, breathing naturally, also continue to feel the slight tingle of the breath as you breathe in and out. If you find that your attention has wandered, pull it back to the tickle of breath. You will find yourself becoming very calm and focused. Your mind will eventually quiet down until there is nothing in its awareness except the tickle of breath. You will find that your breathing becomes more and more faint as you enter a state of deep concentration. You may find that your breathing stops entirely as you are suspended in a state of extremely deep concentration. Don't worry: as soon as you notice that your breath has stopped, it will have started up again! Do not try to manipulate your breath. *Just notice it.* Ignore any visions, emotions, feelings or realizations that you may have. They will merely distract you from this state of deep concentration. Do this exercise for at least three minutes to start. Then increase your time to seven minutes, then eleven minutes, then fifteen, then thirty minutes at a sitting. If you like, you can sit for an hour. Possibly, when you first start doing this exercise, your mind may seem to be more active. This mind activity has usually been going on all the time, you may have just not noticed it before. Most importantly, do not judge yourself or your "progress." That will get in your way. Just pull your mind away from your evaluations, back to the breath. This will develop your ability to focus and effortlessly direct your mind.

As you do this exercise, notice how thoughts seem to come and go through your mind. They don't seem to originate in the mind, but seem to float through it. Some are attracted to you, and some are not. Those thoughts for which you have no attraction, or for which you have no tendency to resonate in harmony, do not stay with you. They leave quickly or are unnoticed. In a clear, directed mind, thoughts do not stay unbidden. The thoughts that are specifically called up and willfully directed are those that the clear mind works with. Here is something you can do to practice the process of calling up and directing thoughts and images with a quartz crystal.

EXERCISE TO TRANSMIT PEACEFULNESS

Work with a person who has given you permission to do so. Sit quietly in a relaxed manner with the spine straight opposite each other. Both of you breathe naturally. Have the person opposite you close their eyes. The person opposite you should be still and receptive to what you will transmit with your crystal. Hold your crystal with both hands in front of you, tip pointing away from you, and close your eyes. Imagine feeling very content. If you like, imagine yourself being in the most peaceful place that you have experienced or that you can imagine. Maybe you are lying in the sun on a deserted beach with nothing in particular to do but be very aware of the warmth soaking into you and the cool breeze lightly wafting through the palm trees above you. You may feel like taking several deep breaths and exhaling them, relaxing even further as you do so. Begin to feel a deep contentment within you. Everything is all right. Everything is how it should be, and will always be all right. Imagine that state, creating the feeling inside of you.

Now, as you hold that peaceful image and set of thoughts steady, open your eyes and gaze into the quartz crystal. Still holding the peaceful thoughts and state of being steadily, take a deep breath and pretend to blow all of that peace and contentment into the crystal. Do this several times if you like, until you feel that you have done it enough. Now, holding the crystal in your right hand, point the tip towards the person in front of you who is sitting in a receptive mode with their eyes closed in front of you. As you point your crystal toward that person, imagine all of the peace and contentment that you "blew" inside of the crystal starts leaving from the tip and entering into the person in front of you. This may "look" like a golden or pink stream of light leaving from the crystal's tip, entering the person. As you continue to direct the images, thoughts and feelings of peace into the crystal, they continue to enter the person in front of you. You may feel like moving your crystal around the person to direct the peace to surround them like a halo of light, or you may direct it into their heart. Let your feelings direct you. If your thoughts wander at any time, just pull them back to the process that you are doing. Imagine again the contentment and peace flowing from you, through the crystal into the person in front of you.

If you feel drawn to direct this flow to any particular part of their body, do so. If you intuitively feel any tightness or resistance to the peaceful flow from the person in front of you, direct the crystal to it. Imagine any tightness softening and breaking apart until it is in harmony with the contentment. You may want to imagine the person in front of you expanding slightly with the peaceful glow that is now seeming to emanate from them, or imagine any number of things happening with them that seem to signify their growing contentment. Don't worry, don't wonder if you are doing it right. Just keep doing this process until you feel that you have done enough. You will have a clear feeling of this. Trust yourself. When you have finished transmitting this from your crystal, lay your crystal down in front of you. Close your eyes and sit for a few minutes enjoying this contentment. When you feel like it, open your eyes and instruct the person in front of you that you have finished the process and that they may open their eyes when they feel like it. When they have opened their eyes, sit together for a few moments before getting up and going about your daily business.

Certain things will happen during the process just described. As you imagined feeling peaceful, you created a particular resonance inside of you reflective of this state. The quartz crystal then began to resonate with this vibration as you "blew" into it. As the crystal began to amplify this peaceful vibration, it helped you to feel even more peaceful. This, in turn, fed back into the crystal which amplified the peaceful state still further, which helped you feel more content, etc. As you imagined the peaceful vibrations leaving the crystal to enter the other person, you used your will to direct the vibrations of that person and the crystal to harmonize. The person began being influenced by it and began feeling content.

As you were implementing this process, you might have been aware of one or several things happening to you. Your breathing might have become very still. You might have experienced a tingling sensation at your brow point in the center of your forehead or along your spine. You might have begun to shiver or to perspire. You might have begun feeling weak or charged with unusual strength or energy. You might have felt a heightened sensitivity or awareness in your body. Or you might have begun feeling sleepy. These sensations could be the result of certain energies of a more subtle nature awakening in you. Or they might be manifestations of blockages you have, or more simply, just a weakness of the nervous system as higher, more charged energies course through your body. There are many possible reasons for the various sensations you might have experienced, as will be explained in later chapters. While you do the process, simply ignore these sensations. Drop your attention from them and bring it back to what you are doing. Do not let them interfere with your present focus. You can address and work with them later.

There are two ways of visualizing and creating change. One way is to imagine a change as already having happened and then send that out from your crystal. The other way is the method that was used during this technique. A change is imagined and experienced as happening within you, and then imagined as happening *to its completion* in another person. With this latter process, the important thing to remember is not to force change to happen, but to *allow* it to happen.

The entire universe is a state of mind . . . Change your mind, change your universe

THE WILL

Your will is that determined strength which lies behind your intentions and empowers them. Strength of will is essential in effective quartz crystal work. The length of your concentration, the power of your projection, the clarity of your visualizations all depend on your strength of will. Your will combined with certain techniques creates the many different vibrations in the crystal. Your will also then directs or transmits the energy currents that you have created with the crystal. It is what powers your body and actions on the astral, mental, and causal planes. Your will prevails in all universes.

Strength of will is tied to your overall vitality. If you have little vital energy, you will have little will power. Take care of your health. It is also tied to nervous strength, so it is important to develop a strong nervous system. Also, the more stimulated and open the navel point is, the stronger your force of will. Do the exercises needed to strengthen your navel point. (See page 68 on third chakra exercise.)

Besides your individual force of will, there is another similar force in the universe that is used in effective crystal work. This current of energy resides on all planes. This force is responsible for a certain order, a way things fit together. It is pro-life and one of growth. It is positive. It is harmonious. As you become more conscious, you will develop a sense of this force. It is felt as a purposeful intention underlying all life. We are never apart from it, even when we have no sense of it. However, it includes our individuality and, at the same time, is beyond us. This is sometimes spoken of as a higher will. When you become aware of it, it is not felt as oppressive, but as liberating. Though you can feel it in your body, and sense it, it has no particular form that you can describe and does not spring from any form. It is an essence that exists on all planes.

In your crystal work, as in all metaphysical work, when you are in service to this higher will, you feel harmonious and uplifted. You feel as if you have direction in your work and in your life. A might and strength flows through you.

If you are not in harmony with or in service to this higher will, your efforts will be wasted in the long run. Your results will not be lasting. You won't bring lasting peace and harmony into your life or to anyone's with whom you work. Because you won't receive the direction that you need, you will start relying on your intellectual mind alone to direct you. If you do this, you will be ineffective in the long run. If you actively oppose this higher will, you will pull in negative energy and ultimately suffer.

To be in service of this higher will does not mean that you have no will of your own. On the contrary, what is asked of you demands tremendous strength of will. It takes a strong will to continually be aware of what correct action is asked of you and to do it no matter what. Though your individual will seems to be strengthened when you are in harmony with this higher will, it is still your will that determines your actions. For example, in quartz crystal work, your will changes and directs and transmits the crystal's vibrations. Your will maintains your concentration while visualizing. It is with your will that you direct and hold your attention in all the myriad ways required.

How do you determine what the higher will is and establish your connections with it? You can hear or sense what the higher will is when your mind is still and you are centered. The silent voice of higher will is heard at those times when you listen to your intuition or when you focus silently on your heart center in the center of your chest. It is not the same as your intuitive knowledge, but seems to accompany it in the form of an inner guidance. You can best establish your connection with this inner guidance by asking for it. Make it a daily meditation for a few moments every day to ask for guidance from the higher will. Ask that you may hear and be heard. Ask that your actions spring not from your own will but in harmony with that which is higher.

May that higher guidance or essential order determine your actions. You might try repeating an affirmation silently to yourself as you go about your daily life..."not my will but thine..." Let it permeate your body, mind and spirit until it becomes a part of your very nature.

*Are you searching for your power
and wondering how to use it?
True power is not something to be possessed.*

*You will never find your true power
or find the wisdom to use it
until you are beyond having to be right.
In trying to be right, who are you . . .
or what are you defending anyway?*

*Stop trying to hide from yourself.
By doing that, you create your own prison.*

PRANA OR LIFE FORCE

Prana is referred to as the life force of the universe. It is found on every plane and vitalizes every being, ourselves included. This vital force sends its life currents down through all our bodies to the physical where it is carried by the subtle nervous system and absorbed into the cells themselves. Your degree of health and vitality is determined by your absorption and circulation of prana. The more prana you absorb into yourself, the more vitality you will feel. When you have more vitality, you are healthier.

Prana is intimately tied to your breath. When you deeply inhale, you pull more prana into your body. Similarly, when you exhale, you are discharging it. You use your breath to circulate this pranic life current in your body. This distribution happens naturally as you breathe. However, prana distribution can also be directed with your will inside as well as outside of your body. There are many techniques using your breath to distribute and collect prana in your body. These are called Pranayam.

You can also use your will in combination with your breath to send this life force out to vitalize any other person. (This is particularly good in healing work.) You can also use your breath to charge, vitalize, or empower any projection that you may be attempting in your crystal work. In other words, you use your prana-charged breath coupled with your will to carry the energy current released by your crystal. To do this, you consciously collect prana with your inhale. With your exhale, you send this prana out to charge your intention. (You will see this process utilized many times in this book.)

The following is an exercise that you can do to practice sending prana out to vitalize another person. You can also use this technique to send extra prana or life force into your food, your plants or animals. This is best done outside in sunlight, but can be done anywhere.

PRANA EXERCISE

Sit or stand comfortably with your spine straight. Hold a single-terminated quartz crystal or a crystal wand in your right hand, the point outward from your hand. Center yourself as you close your eyes and begin to breathe with long deep breaths through your nose. This is a particular breath technique that, among other things, works to draw extra amounts of prana into the body. Fill your lungs completely as you inhale. As you breathe in, imagine that you are pulling in with your breath large amounts of prana and circulating it throughout your body. As you exhale, imagine that you are sending out from your body any negative energy. Feel as if your body begins to radiate more and more vitality with each in-breath. Do this for at least three minutes. Now, as you continue to inhale prana, visualize the person to whom you would like to send this vital energy. Hold them clearly in your mind's eye. As you continue to focus on the person, hold your right arm and hand up, seeming to point the crystal or wand at them. As you point the crystal or wand towards the person, inhale through your nose. Then exhale through your mouth towards the crystal, filling it with prana from your body that you have accumulated. Do this two or three times, or until you feel that the crystal is charged with this vital force. Then inhale again and exhale through your mouth sending the prana from the crystal to the person. Continue to send the prana force from your body, through the crystal, to the person until it seems like time to stop. Don't do this for more than ten minutes. Then, put your crystal aside and drop your arm to relax by your side. Re-charge yourself with prana by inhaling it in and exhaling any negativity you might have pulled in. Do this for at least three minutes. Then relax.

You can also do this process with the person right in front of you. It does not have to be done from a distance. If you do it this way, continue until the person feels more vitalized or until you see that they are. Do not do this for more than ten minutes at a time. Always be sure to charge yourself with more prana after this process. When you are through, clear your crystals.

WORKING WITH EMOTIONAL STATES

To work with emotional states and quartz crystals, you must be in firm control of the emotions. They must not control you. To be in control of your emotional states does not mean that you don't have emotions. On the contrary, you not only have emotions, but you must develop the ability to feel them intensely. If you do not have the ability to feel your own emotions, you cannot open yourself to feel the emotions of other people. You are then severely limited in the crystal work you are able to do. So much of crystal work demands that you have empathy and compassion with those that you work with. To have these qualities, you must have an open heart center because it is in this center that empathy and compassion dwell. If you harden yourself against feelings, you build walls around your heart center and close it down. If your heart center is closed, the energy that you need to work with cannot channel through your body as it must, and you become blocked.

What does it mean to be in control of your emotions? It means that you allow them to come and go without necessarily letting them determine your speech or your actions. That impartial part of you that can just observe them will then determine appropriate action.

As you work to strengthen your nervous system and open your navel center, you will develop that strength of will that is necessary to be able to withstand the urge to act impulsively. Then you will be able to step back, just observe and experience the emotion without repressing it or in any way doing anything about it. The less involvement and attention you pay to an emotion, the quicker it will disappear. You feed emotions with your attention to them. This is true about any emotion, not just negative emotions. Observe what triggers each emotion. Usually it is based on a desire or set of desires you have that are not met. Experience the emotion and that desire behind it completely without taking any action. Then let go of the desire. You will learn much about yourself and about others.

When effectively working with crystals, you must be flexible and act in accordance with the guiding voice within you. If you are attached to or lost in an emotional state, you will not be able to let it go in order to focus your attention on that inner guiding voice. You also won't be able to use your will to project any other visualization, thought or emotion. So, instead of doing what you intended in your crystal work, you will primarily be projecting your emotional state.

Be cautious and conscious of what you project. Clear your crystals and tools and the room you are/were in if you have been around them in a negative emotional state.

When you are able to just observe your emotional states without acting on them, you will eventually understand how to deal with them. Emotions are not morally right or wrong. The actions that you take based upon them are judged right or wrong. Right action is that which is in harmony with your environment with others and with your own inner voice.

As you continue to develop dispassion in yourself, your heart center will open. As your heart center opens and you are not blinded by your emotions, you will find that you are in a natural state of love. It only has to be uncovered. This natural love has nothing to do with emotionalism. It is much more calm, more expansive, and much deeper.

You will see that every situation, object and being has a particular feeling associated with it that you can feel in your body when you are clear and centered and when your mind is quiet. When you can sense or feel this emotional component, you are said to have empathy with it.

Here is a suggested method to work with emotions: First, deliberately create an emotion in yourself and project it with the help of your crystal to make changes. How is this done? Begin by centering yourself, clearing your mind while holding your quartz crystal. Picture a particular change you want to make or a visualization that you want to project. Then focus on your body in the area of your heart center in the center of your chest. What feeling or emotion seems to be associated with what you want to do? If you don't feel a particular emotion, imagine one. Then while still focusing on the visualization or change, increase the intensity of the emotion that you have in your body. Using your will, project the emotion as well as the visualization into the crystal. This will create a particular vibration in the stone that will correspond to a highly amplified version of what you intend to project. Then while still focusing on that visualization and corresponding emotion, use your will to transmit that vibration from the crystal to the intended situation, object or being. Use your exhaled breath to empower your transmission. In this procedure you have worked with the emotional (astral) and mental plane to influence the physical. To create and transmit emotions, you need to be able to put aside any emotional state that you happen to be involved in at the time and just focus on the emotions that you want to work with. In order to do this, you must develop the ability to control your emotions.

The following is an exercise that will allow you to control your emotions, thoughts and desires:

CRYSTAL EXERCISE TO LET GO OF THOUGHTS AND DESIRES

Sit on a chair with your feet flat on the ground. Or, if you are more comfortable, sit on your knees or cross-legged. Keep your spine straight. Hold a quartz crystal in front of you with both hands. Point the tip upward. (Holding the crystal with both hands balances the male/female energies in your body. This balance helps you channel energy through your body.) Take a few long deep breaths through your nose to center yourself. Focus your concentration into the crystal so that you vibrate in harmony with it. Think of the desire or thought that you would like to eliminate. Feel any emotions associated with that desire and retain them in your focus. If any pictures or visualizations come to you that seem to be associated with this unwanted desire or thought, focus on them also. When you have the thought or desire and all associated with it firmly in focus, take an in-breath through your nose. Then forcefully exhale through your mouth and pretend you are blowing all unwanted emotions and thoughts into the crystal in front of you. Continue this process for eleven minutes or until no unwanted associations come to mind. A vibration in the crystal will be created which resonates in harmony with the projected desire or thought. The crystal might begin to seem dull or cloudy. When you are through, take the crystal with your right hand and point it down into the earth in front of you. If you are inside, point it down to the floor.

Use your will to send the desire or thought from the tip of the crystal into the earth. If you are inside, send it through the floor or floors into the earth. Imagine the particles of the earth surrounding and breaking apart the desire or thought until it is completely disintegrated and swallowed by the earth. Then, imagine the earth to be in a state of calm and peace.

Continue to do this for eleven minutes, until you feel it is time to stop, or until the crystal seems or looks more clear. The vibration of the crystal will have changed back to its prior vibration or to a new vibration matching your current peace of mind. If you like, bury your crystal in the earth for a length of time that seems right to you. Clear the crystal, any other tools, yourself and the environment around you. Place your hands on the earth briefly, then wash them.

Emotional states are the creation of mind (You can choose how to react)

The mind obeys the will
What lies behind the will?
Where is the location of that which lies behind the will?

PROJECTING EMOTIONAL STATES, VISUALIZATION, AND THOUGHTS WITH CRYSTALS

How do you use your will in conjunction with a quartz crystal to project the vibrations you created in it which correspond to an emotional state, a visualization, or a thought? First, set three crystals of equal strength or vibration in a triangle around you with a crystal at each point of the triangle. Sit or stand with your spine straight in the middle. Face the top point. It is out of this point that you will send your projection. Sensitize your hands. When your hands are sensitized, hold a quartz crystal wand or single-terminated crystal in each hand. The two crystals should be matched in size and power. (You can visualize the crystals if you don't have them.) Close your eyes and begin to concentrate.

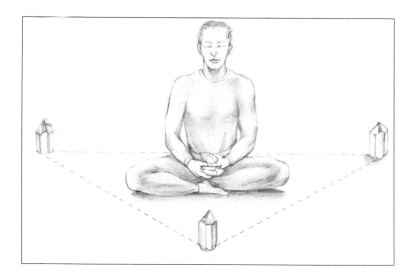

Take a few long deep breaths through your nose to further center yourself. As you do this, focus on your heart center in the middle of your chest. As you inhale, fill your heart center with a loving feeling. As you exhale, send out anything other than that love. Continue until you feel as if you are completely filled with this loving feeling. This will activate your heart center. Next, extend your arms to your sides, parallel to the ground. Hold your sensitized hands with their palms up, holding the quartz crystal(s) or the wand. Point them in toward your body.

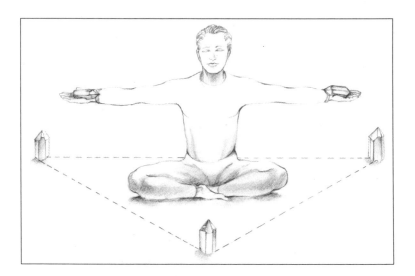

As you inhale through your nose, feel that you are drawing in currents of energy through your fingers. Imagine this activating the crystals, amplifying the energy even more. As you send this energy up through your arms to your heart center, mix these powerful vibrations inside with the love already there. Send them through your entire body. Continue to do this until your entire body is filled. Use long, deep breaths, and inhale through your nose. When you feel yourself filled with this vibration, focus on the emotion, thought and/or visualization that you want to project. Focus on it very clearly. See and experience every detail. Fill your body with this charged vibration. Continue until there is no other thought, emotion or experience of your body that is not part of your intended projection. When that state is reached, send the vibrations from the heart through your throat center and up to your third eye or the energy center in the middle of your forehead.

Focus on your third eye point as you turn the crystals or wand to point outward from your body, reversing the energy flow. Hold them in front of you with your arms extended, or hold only the right arm out.

Now, inhale a deep breath through the nose and blow it strongly out through the mouth. On each out breath, send streams of energy out through your arms, hands, third eye and crystals. With the outward current of energy, send the thought or visualization to the intended person, being or object while focusing on your third eye.

Constantly fuel your third eye with the energy from your heart center. (This is only possible if the throat center is not blocked.) Fill the other person with the projection until your body feels empty, until you see results, or until you feel naturally it is time to stop.

What do you use to project? You use your will, your strong intention. It is as if you empower or push the vibrations out with this intention. The strongest will is backed by faith or belief in what you are doing *in the exact moments that you do it.* It is possible to base your belief on what you have heard or read about. However, for most effectiveness, base your belief on your own experience.

When you are finished, sit for a while and center yourself, gently filling yourself with energy again. Don't leave your body and its systems depleted. Draw in energy from the earth through your feet to nourish you or draw in from the sun through the crown of your head. Then clear the room, clear yourself, and clear your tools.

Only use this technique when asked, or in some cases when you feel a strong inner guidance to do it. Before you change things in the physical universe around you, contemplate the action and the ramifications of the change. Is it the proper thing for you to do? Is it in harmony? Are there good reasons for the situation to remain as it is? As you contemplate, "view" the situation through the perspective of both your heart and your third eye centers.

Be aware of the overall plan of this life on all planes, not just the physical. Don't create chaos. Don't be indiscriminate. To be able to view the overall plan, develop yourself through the suggested techniques in this book. Act using the viewpoint of the higher spirit or the higher, more subtle planes.

The more attuned your work is with the overall plan, the more you can act as a "clear channel" letting energy flow through you without impediment. You need to strengthen your will, open your third eye point, your heart, and other energy centers so that energy is able to flow through you. (See later chapters for methods to do this.)

Be willing not to use your crystals or do any work at all unless you are called to. Be perfectly content either way. If you are content to do nothing, you are less inclined to do crystal work when it is not needed. Don't do any of this work to show off because this invariably proves harmful to yourself as well as others. It is a trap that will pull you off center and you will not be able to hear your inner guidance. When you are off center, you don't feel the joy and deep contentment that you will feel if you are acting merely as a conduit so that the natural flow of harmony may do its work.

2

The Extended Body

THE HUMAN BODY has surrounding it ten other bodies, each in its essence, vibration. As you will see in the following discussion, in quartz crystal work you are chiefly concerned with four of them: the physical, etheric, astral, and mental body. The different bodies can be pictured as layers, one on top of each other even though in actuality each body is contained within the

47

subsequent higher bodies. They surround the physical body roughly in the shape of an ovoid. (See illustration.) Each has its particular density and type of vibration. As the bodies extend out from the core, or physical body, they become increasingly more fine in vibration. Though all of these bodies are of different densities, they correspond and are joined to one another. Thus a change in one body affects all of the others.

When you create a change on a subtle vibrational level to affect the physical body, as explained in the beginning chapters, you are actually working directly with the vibrations of the etheric body, the astral, sometimes the mental body and very occasionally the causal body. As will be shown later, you can work solely with one body or work with several at once. This elucidation of the various bodies and subtle planes is, however, meant to give you an intellectual context within which to work. Do not use your intellectual understanding of the extended bodies and planes to replace actual experience. It is not necessary even to know about these to do effective work as long as you can feel the vibration and are guided by that feeling. Just do your work using this knowledge as a framework to allow you to open to new possibilities of experience with your crystals.

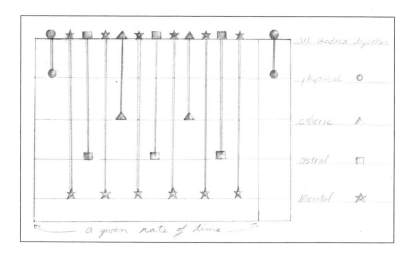

The physical body has the slowest rate of vibration among all of these bodies. This physical rate of vibration is represented above by the round shapes. (○) This vibratory rate represents a type of mass which creates our physical body and the senses of which we are the most aware. (There are bodies more dense than our physical body but these are generally irrelevant to our work and are not mentioned here.) The etheric body vibrates more rapidly than the physical body, but not as rapidly as the astral body. To vibrate more rapidly is referred to as being "more fine." Because the rate of vibration is more rapid, or finer, more of the vibrations will fit in a given rate of time in the same space. This is illustrated in the chart above by the triangle shape (Δ) representing the etheric body. Next in fineness is the astral body represented by the square shape (□) in the chart. Notice how the astral body, because of the rapidity of its vibration, is contained within the etheric as well as the physical body. (The shape is contained within the set of ○ and Δ shapes. The astral body extends beyond the physical and slightly beyond the etheric body. Likewise, the mental body is higher in vibration than the astral, etheric, and physical. You can see in the chart that the mental body's vibration, represented by the ★ shape, fits in between all the □'s, Δ's, and ○'s due to its higher rate of occurrance in the set time space than the other, slower vibratory rates. (There are higher bodies that vibrate still more rapidly. However, we don't consciously use these during crystal work so they are not shown on the chart.) If you look at the chart you can see that changing one rate of vibration affects all the others since they are interlocked with each other. A change in one body affects the others.

Therefore, in quartz crystal work, we can work on a subtle body to affect the physical body.

You also can see that the term "astral travel" and "out of body travel" can be misleading. When you are in these states of consciousness the sensation that you experience is rather like one of travel or movement in space. However, contrary to the "travel" sensation, you really go nowhere because the astral is contained within the physical body. Rather than traveling somewhere, you expand your awareness, and your focus. You become conscious of different impressions, those of the astral plane. Then, like the physical senses, they become the context from which you operate. This is the main method of working astrally with crystals. Astral travel and out of body travel refer to maintaining your conscious awareness in your astral or other body, and traveling using that body. For those with this higher awareness, it is possible to travel to your new astral or other place of location and then also appear in the corresponding location on the physical plane. It is very rare for most crystal workers to do this. Most just travel on the astral plane in an astral body and then return to the physical body in the same location from which they began. However, you can see how this could be done knowing the true interrelationship of the various subtle and physical planes and bodies.

Now that you have seen how the various bodies and universes interrelate, what are their unique characteristics? First, it must be remembered that bodies are essentially vibration. Each body of a particular plane is a conglomerate of the general vibrations of that plane held together by a particular force of consciousness. (This is also true of the physical plane.) That force of consciousness regards itself with separate boundaries, edges, or characteristics that set it apart from what is around it. It is like being a subset of a larger set. Thus, the force of consciousness gathers a body around it in a particular plane, having all the characteristics of that plane because it is the same as it. In other words, a body is just like a "piece" of a world. To remain a "piece" of a world, a body must retain some sense of separateness from it. So the body or "individual" is not aware of all that it shares with its universe or it would no longer be able to experience its separateness. This experience of separateness does not reduce the shared characteristics with the corresponding plane but only inhibits the use of those characteristics due to the lack of focus on them. That is why you are not usually immediately conscious of all the subtle-plane characteristics that are contained within you and have to first let go of some deep-rooted assumptions of what is "real" or not.

This has been explained to you for a number of reasons: First, contemplating this process and seeing much of it in yourself will begin to open your awareness. As your awareness expands you will begin to be conscious on other planes beside this physical one. This will be helpful not only in terms of general enlightenment but will give you more ideas about how you might do your crystal and other metaphysical or healing work. Also, if in doing this work or any meditations or other practices, you begin to have out of body experiences you won't be scared. You will have a context by which to understand and make use of this experience.

In the next sections the characteristics of each plane will be elaborated upon. The descriptions will be mainly about the plane itself. As explained earlier, remember that the description of the physical plane extends to its corresponding body. Any other specific characteristics of the body that are important will be elucidated also.

When you effortlessly and continuously dwell in and act from this consciousness you will be able to work perfectly with quartz crystals...or anything.

Ultimately you are just consciousness itself When you effortlessly and continuously dwell in and act from this consciousness you will be able to work perfectly with quartz crystals . . . or anything.

THE ETHERIC BODY

The etheric body, easily confused with the astral, projects about one-quarter inch beyond the skin. Its aura can project several inches beyond it, usually no more than a foot. However, with certain practices, increased vitality, and/or placement of crystals on the physical body, the aura can extend at least another foot and sometimes several feet. The etheric body is a perfect duplicate of the physical body. This body is not a separate vehicle of consciousness from the physical body. However, it is completely necessary to its life. The etheric body receives and distributes the vital forces that emanate from the sun and so is vital to the physical health. The vital forces with which this body is concerned are the kundalini energy and prana or vital force. (Kundalini and prana are not exclusive of this body, however, but are known to affect all the others.) Also the etheric body works with practically all of the most familiar physical forces, i.e., magnetic energy, light, heat, sound, chemical attraction and repulsion, and motion.

Prana or vitality is collected, filtered and distributed by the etheric chakras or energy centers through the body in various energy channels or pathways to be the controlling energy working through your nerve centers. This keeps both the etheric and physical body alive. The etheric body acts as a two-way bridge between the physical and the astral. The etheric chakras bring into the physical consciousness whatever the inherent quality is in the corresponding astral centers. With this bridge your dreams are brought to consciousness when waking. Also, through this bridge, your physical sense-contacts are transmitted through the etheric "brain" to the astral body. Similarly, your consciousness from your astral body and other higher bodies is transmitted into your physical brain and nervous system. As you begin to automatically expand your awareness with crystal work, these subtle systems awaken and increase. As you can see, it is with this body and its systems that you so intimately work with your quartz crystals. Therefore, kundalini energy, and the chakra system will now be explained more in depth. Exercises will be given to develop your abilities to experience and consciously work with these energy systems. This will tremendously improve your crystal work, opening up vast areas of possibilities.

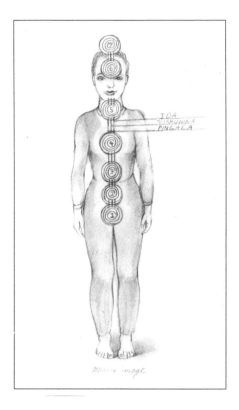

CHAKRAS, IDA & PINGALA, SUSHUMNA

In the etheric body are seven energy centers which also have their astral counterparts. These are called chakras, which means a wheel or revolving disk. They appear to be rapidly rotating circular shaped vortices of energy or subtle matter. The chakras are points of energy transformation, absorbing, filtering and distributing vitality to the etheric as well as the physical body. As the chakras are stimulated and opened they transform energy frequencies, bringing into physical consciousness the inherent qualities associated with each one. This results in different behaviors, different states of health, and different levels of awareness or consciousness.

There is a central energy channel running up the subtle body connecting all the chakras from the first to the seventh or crown chakra. This is called the sushumna. This channel roughly corresponds to the spinal column in the physical body. Along the left side of the sushumna is a channel of feminine, or moon energy called the ida, receptive in nature. To the right of the sushumna is the pingala, a channel of masculine, more outward sun energy.

Notice that all bodies have masculine and feminine energies in them. This refers to the naturally circular flow of energy that both receives as well as manifests. This does not refer to your masculinity or femininity which are by and large social labels attached to certain forms of behavior.

In quartz crystal work we work directly with the ida, pingala, sushumna and chakras: by affecting them we immediately affect the entire physical, mental and emotional bodies. Most illnesses and other problems are the result of some blockage or imbalance of the chakras, the male/female energy flow or the central cord of energy. As you become more aware, you can see or sense where the imbalances lie and correct them. As you experiment with the chakras and become familiar with their qualities, you will be able to stimulate or open them with your crystals, bringing those qualities into the body of the person that you work with, (including yourself). Usually people do not have any of the higher centers open, so that the qualities associated with each center are unknown to them. To correct this situation, you will want to open the higher centers from the heart chakra up. Use the heart as the center or the point of balance. Do this gently, always using your intuitive senses or inner voice to let you know how much is good for someone.

As you open all the chakras with your crystals you also need to be aware enough to keep them balanced with one another. Be responsible as you work. Do not try to direct an energy flow into the chakras unless you can see, feel, or sense that flow. Don't rely only on intellectual understanding to do this. You must know what you are doing, using your direct experience, or you can unbalance yourself and/or others and hinder rather than help. To get to know each center you can sit quietly and meditatively and focus on it. Mentally record any impressions, feelings and other information that you receive. The following is a discussion of the seven centers, their location and the attributes associated with each one. Following that are practices that will allow you to open each center. Experiment with these yourself before you use them with others. The later chapter on healing has more information about working with the various energy chakras.

DESCRIPTION OF THE CHAKRAS

There are many varying descriptions of the chakra systems, both within ancient yogic and tantric texts and in the more modern accounts. Rather than burden you with the myriad of information available, a more simplified exposition of this system is presented here. (See chart.) It must be experiential to be of any use to you. First, learn its basic structure. Then use your crystals to meditate on and work with this structure. As you do this, more and more of the complete system will be intuitively revealed to you. While keeping the basic etheric energy system in mind, if you intuitively decide to add to it or vary from it, do so. Check your results. Be in the moment. Don't blindly insist that certain pathways or relationships must always be directed in a certain way. That might not be the case. Each person, yourself included, is in a different degree of psychic, physical, emotional and mental health and evolution. As you work with each person, your job is to notice what parts of the subtle energy system seem to predominate and/or need treating at each moment that you work. If you have memorized an elaborate system that you are not also experientially aware of you can become trapped in mere dogma. You will be unable to do spontaneously effective work. The system presented here is one that most of the various texts seem to agree on and one that you will easily be able to work with.

The first chakra lies in the vicinity of the bottom of the spine. The ancient name for this is called Muladhara. This is where the kundalini resides and is sometimes referred to as the "seat of the kundalini." This center is mainly concerned with basic survival. The color associated with this center is red. Most of the time when any crystal work is done with this center it is either because there is a specific physical disease centered there that needs healing, or work is being done to stimulate and raise the kundalini force, or the person needs to be grounded. Other than that, people usually are very centered there already and do not need any more stimulation in this area.

The second chakra is called Manipura or the spleen, or sex center, depending on what texts you view. Its location roughly corresponds to the area of the sex organs. The color usually associated with this chakra is orange. The second chakra is concerned with the sexual urge and creation on a less subtle level than the other centers. This center, like the first chakra is usually open in most people. The times this center needs to be directly stimulated are the following: when there is a physical disease centered there, sexual blockages or disfunctions, and when arousing the rise of the kundalini from the first chakra through this center on to the third chakra. It is also stimulated during some forms of tantric yoga.

The third chakra is called the navel center or svadhisthana. It is roughly located in the area of the navel, about two inches below it. This chakra is concerned with vitalizing both the physical and etheric bodies. It is where all of the 72,000 subtle nerves join to filter, transform and distribute subtle energy throughout all the subtle body. The third chakra is the seat of the will which empowers and helps manifest the unmanifest impressions from the higher centers. Use of your will powers your action on all planes. If this center is open and is not balanced with the higher centers, people can become overly concerned with issues of power and control.

The navel center is the source of energy or the center for physical well-being. All imbalances of the mind or alterations of human behavior reflect some imbalance in the lower three centers. The navel chakra can transform these imbalances when it is opened and strengthened. This center gives you the ability to create and break habits and to hold strong to a determined course of action. If the navel center is weak and unstimulated, certain physical and psychological manifestations can appear. Physical symptoms can include premature old age, lack of nerve strength, and in later life, brain and organ failure and cancer. Psychological symptoms result if this center is blocked from transforming energy to the heart center. A person will show greed in all aspects of the personality: for example, compassion and other human values are shown only if they will secure further recognition, selfish ego enhancement, etc.

The color associated with the third chakra is yellow, sometimes pictured as a fiery yellow sun. When doing your crystal work you need to be cautious when stimulating this center. Most people have the third chakra open to some degree, but it may not be balanced with the open heart qualities or the vision of the open third eye. If you open it more without also working with the heart and/or third eye you can lock people even more into a position of having to have power and control over other people and the environment around them. This produces much unhappiness and discontent. Sometimes it is enough to just open the higher centers, and the will power necessary for crystal and metaphysical work will be there along with a lessening of the psychological need for power over people. Development of this center helps to attune you to the astral plane. It also helps you to manifest what you have seen on the subtle planes to the physical plane. The sound associated with this center is HUH (as in the sound butter). The mantra associated with it is HARA. (See exercise for opening the third chakra, pg.68)

The next chakra, or the fourth chakra, is called Anahata or the heart center. It corresponds to the area in the center of the chest, between the two nipples. The color of the fourth chakra is green, although many times in crystal work a rose pink color is used. Sometimes rose pink is considered to be tied to the emotional aspect of the heart center, while green represents the chakra itself. As with all crystal work, use your discretion. The sound associated with it is "AH" as in "mom." A mantra often associated with this center is RAMA. In its lower sense, this is the center of emotion. If someone is having trouble with being over emotional, work with the heart center is called for. If a person is feeling "drowned" in sadness, for example, you might use a pink stone to massage this center or you might use a green stone to strengthen the heart center. As you do this, you might want to open their third eye center and connect it with their heart center to help them channel some of the excess "sad" energy upward. In the higher sense, the fourth chakra is associated with love and compassion. When this center is activated you become aware of the emotional feelings of others and extremely empathetic with all beings, instinctively understanding them. The heart chakra is the place of centeredness, of balance. It is the joining point of the etheric energy flow between the earth and sky that is used in healing. (See illustration, page 60).

The fifth chakra is the throat center. The ancient term for it is visuddha. It is located in the etheric body in a location that corresponds to the area in the center of the throat. The color associated with this is a light or sky blue or turquoise. The sound associated with this chakra is 000 or HU (as in who). When the throat chakra is open you gain the ability to hear and bring to physical consciousness sounds from the subtle etheric and astral planes. This is called clairaudience. You gain the ability to hear truth from the higher planes, and to communicate it on the physical plane. When this center is open that you can hear what to do with your stones. As your throat center opens and you begin to speak of what you hear, listen carefully to know when it is appropriate to speak. Do not speak when people are not in the position to hear. This often takes much patience. The throat chakra is the center through which the heart connects with the head (or the third eye). To be balanced, this connection needs to be made. Headaches or tightness in the jaw, shoulders and/or neck tension are often signs that the throat chakra is blocked. In your work, compassion and wisdom need to be united, then communicated.

The next chakra is the ajna or the sixth chakra. This is sometimes called the third eye. The sixth chakra corresponds to the point between the brows in the center of the forehead. It is the development or opening of this center that brings about the clairvoyant abilities. As the third eye opens your intuitive powers develop to their fullest degree. Sometimes you may start having visions of fantastic colors, people and places. This is the center of astral sight. You start seeing clearly the cause and effect relationships of events and people. You develop a sense of the ultimte perfection of life, even as you also see the need to help change that which seems imperfect on another level. You develop wisdom.

The open third eye sends you into a state of mystical all-seeing, all-knowing awareness. This joyful experience is beyond all description and all rational understanding. In this state, you will naturally know what to do with your crystals. Focus on what you would like to do with the stones and the mechanism will be revealed to you. Be careful as you open the third eye. If you stimulate too much energy in this area without balancing it with the other centers, you can create painful energy blockages. Also, you might have experiences that you are unable to integrate using the "wisdom" of the other centers. It might even become very difficult to operate in everyday life. Sometimes as you work you may need to close down this center slightly and work to open the heart and throat centers more. As with all crystal work, *balance is the key !*

The crown chakra is located in the center of the top of the head in women. Men have this center located slightly more forward of center. It sometimes feels as if it's hovering slightly over the top of the head. Its ancient name is Sahasrara, and is sometimes called the center of a thousand petals. Its color is purple, even though when you are focused on the crown center you often seem to be awash in golden light. You can best visualize this as a large sunburst over the top of your head. It seems to continue indefinitely as you focus on it. The sound that is associated with it is "MMM" and as you make that sound you hover it over the top of your head. The other mantra or sound you can use is "AUM" or "OM." As this center opens you seem to become one with the universe. It doesn't feel as if you are leaving your body: you just have a feeling of being one, not separate from anything. There is nothing to be separate from. It's a feeling of ecstacy, gentle ecstacy. There are no words to adequately describe this state of consciousness.

The crown chakra needs to be open during crystal healing work or any healing work you do. It's the center through which you gain your highest wisdom and highest knowledge. This knowledge filters down through the other centers as it begins to manifest in some sort of form. The crown center should also be open so that the energy that you work with can come through the top of your head, filter through the upper centers and meet in your heart: then it flows out in every way that it should manifest, using the different centers. You also will have the bottoms of your feet open and the bottom chakra which connects you with the earth. Join this energy with the upper chakra energy in the heart. (This energy flow is covered in the healing chapter.) You now have a continual circling from the earth to the top of your head and from the top of the head into the earth. This two-way flow is symbolized by the double triangles: the seal of Solomon and the Jewish star. One triangle reaches up, the other reaches down. The dorje is like this: Energy comes in one tip and out the other tip, and in that tip and out the opposite tip. A double-ended crystal also creates this two-way flow of energy.

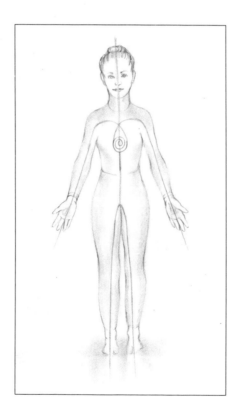

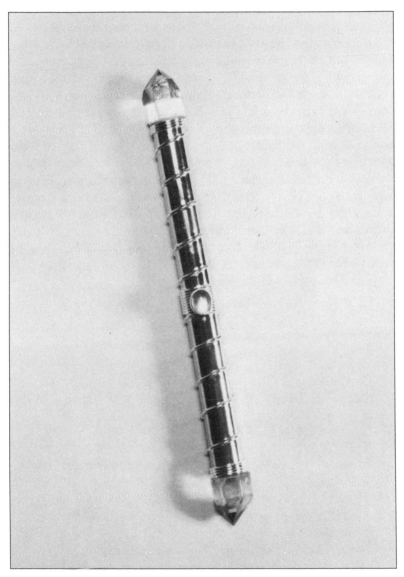

Quartz crystal dorje to work with two-way energy flow. Sterling silver and copper.

With the crown chakra, the two-way passage of energy is established as the center opens. In newborn babies it is very open. (The soft spot on the head is right where the crown center is.)

Here is a meditation you can do to learn to experience that two-way flow of energy. This will help channel any of the information you should need to do the quartz crystal work. To do this exercise, you can use either a double-ended crystal or a crystal dorje. A crystal dorje is like a wand with a crystal in each end pointing in opposite directions. It is a replica of the two-way energy flow in the subtle body. Even better, use a crystal dorje with one clear and one smoky crystal. The clear corresponds to the crown center, and the smoky with the first chakra. Hold it level with your heart. (See diagram.) If you don't have a dorje, use a clear crystal over the head pointing out from the crown chakra, and a smoky near your tailbone pointing down or under your feet pointing down into the ground, creating a dorje out of your body. (This can best be done when lying on your back.) A dorje tool is a replica of that energy flow.

EXERCISE TO OPEN TWO-WAY ENERGY FLOW

Sit, stand or lie on your back with your spine straight. Center and ground yourself. Now, breathe in and out a few long deep breaths and concentrate. Breathe through your nose.

On the next inhale, imagine breathing the ground in through the bottom of your spine, then into the heart. Let it circle in your heart center in a clockwise direction. Let it circle a couple of times and then send it out the top of your head on the exhale. Continue to do this for at least one minute. Finally, inhale from the ground, mix it in the heart, exhale from the top of the head, and feel the stream from the top of the head continue, with no stop.

Now, similarly, begin to inhale through the top of the head, down through the body into the heart. Circle the energy a couple of times clockwise in the heart. Then exhale, sending it out through the tailbone into the earth. Continue to breathe this way for at least a minute or as long as you did the first part. Now take a deep breath through the nose. Inhale through the top of the head, mix it and circle it in the heart and send it down through the tailbone into the ground with no stop. Imagine it going down into the earth with no end.

Now breathe in through the bottom of the spine, up the body, circle in the heart and exhale, sending it out through the head. Then reverse it. Breathe energy into the head, circle it in the heart and exhale it out through the first chakra. Continue this process of reversing your breath. The dorje or double-ended crystal you are holding level with your heart center helps keep you balanced. Now experience turning the dorje in the other direction with the smoky crystal pointing up toward the top of the head and the clear crystal pointing down. Or put the smoky crystal over your head and the clear one under your feet or pointing down from your tail bone. Continue your breathing in one direction and then another. This is like doing a headstand or shoulder stand, reversing the flow of energy. This exercise is very balancing, healing and energizing.

There are some other chakras or energy meridian points which haven't yet been discussed, some of which we use and some of which we don't often use. The first and most obvious are the energy points on the soles of the feet. When we talk about grounding from the feet, or being attached to the earth, or energy coming in from the earth, it is through this meridian point. Imagine a place or an opening right in the middle of the sole of your foot. That is where this meridian point is.

Other chakras are in the center of the palms of the hands. In ancient tests, this is sometimes shown as an eye in the center of the palm. You can channel energy from any other chakra point through the palms of the hand. The healing energy flow also can pass through these meridians.

There are more chakra points than those I've mentioned and it's good to be aware of them. But you don't need to work with them on a conscious level for crystal work. There is a chakra or energy vortex about six feet below us in the earth. There are others further over the crown. These are the eighth, ninth, and tenth chakras. If you choose to familiarize yourself with these upper chakras, first open the crown center and then focus up higher, about three to six feet. Then keep focusing higher. Information about them or a sense of them will come to you.

In quartz crystal work there is not a particular need to work with these upper centers. You can do everything you like working with the simple seven chakra system.

There is an important point to consider when working with the lower chakras. Most people have these open to some extent, as noted above. However, some people have these lower centers entirely blocked. This often is the case when the person considers the lower three chakras to be "unspiritual" or bad. They then try to ignore these centers or consciously close them down and try to dwell only in the upper chakras. Often this person is undergrounded or "spaced out," mistakingly placing a moral judgment on the terms "higher" or "lower" plane. Higher does not mean better and lower does not mean bad or worse. These terms refer only to the relative position of each energy center. Your lower chakras help connect you with everyday life. You cannot manifest what you see and learn on the higher planes without your connection to this everyday life.

In working with people you must be conscious of their purpose. If a person wants to be more spiritual or wants to learn to work with metaphysics, for example, they will generally need to deflect their attention somewhat from the lower chakras and focus more on opening the higher centers. Also, there is a genuine spiritual path where the sexual energy is turned upward to the higher centers to raise the kundalini. This often tends to close the lower centers. If this is the person's path when you work with them, you would not want to interfere by opening the lower centers. At some point, however, the lower centers have to be open so that all centers are balanced with each other. Again, as you work with yourself and others, be conscious and work with much discrimination. Each person and each situation is unique. The Cassette tape "Crystals, Chakras, Color and Sound" will further assist you to work with chakras and subtle energy channels through the use of sound and visualization.*

EXERCISE TO BALANCE THE MALE/FEMALE ENERGIES IN THE BODY AND OPEN THE HEART CENTER

The receptive female/moon energy flows downwards through the ida channel. This is on the left of the sushumna. The manifesting male/sun energy flows upward through the pingala, the channel on the right of the sushumna. These are non-physical, existing in your subtle body. The ida and pingala flow of energy must be balanced with each other in order for you to receive and send energy through your body in your crystal work. The following exercise balances the male/female energies, allowing them to flow freely. It also opens the heart center.

The sound RA, pronounced "Rah," refers to the male, solar energy. The sound MA, pronounced as in 'mom,' refers to the female lunar energy. The seed sound AH which is included in each stimulates the heart center.

---

*Crystals, Chakras, Color & Sound, ©1986 Uma and Ramana Das, U-Music, U-104, P.O. Box 31131, San Francisco, CA 94131

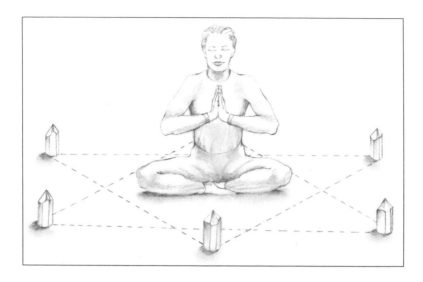

Sit upright in a chair with your feet flat on the ground. Or sit cross-legged or on your knees on the floor. Keep your spine straight. Hold your hands with the palms and extended fingers flat against each other as if you were saying a prayer. Press them firmly against your heart center as they retain the position. This hand position begins to balance the sun/moon energies. Close your eyes and focus on your heart center.

Breathe in and out of your nose with long deep breaths, completely filling and emptying your lungs. As you do this, feel your mind begin to calm and your concentration effortlessly increase. When you feel centered inhale, and while holding your breath sing RA for four seconds, then MA for four seconds. Exhale. Continue to repeat this with a steady continuous rhythm. Choose a tone that seems to create a buzzing sensation in your heart center. Then sing both RA and MA as if the sound is coming in and out of that center. Breathe as if you are breathing in and out of your heart. Be sure to maintain a steady, firm pressure with your hands.

Do this for at least three minutes. When you are used to doing it for this amount of time, do it for seven minutes, then eleven minutes. If you like, you can extend the time to a half hour, then an hour. When you are through with the exercise sit quietly for a few moments, breathing naturally. For the full effect, do this for 30 days.

If you would like to amplify the effects of this exercise, surround yourself with a crystal matrix in the shape of a six-pointed star. Use six like-size crystals, one for each point. Sit in the center of the star. Wear a crystal necklace with the crystal over your heart center during the exercise. If you are doing this for a number of days, wear the necklace continually, day and night, clearing it when needed.

EXERCISES TO OPEN THE CHAKRAS
- TO DO FOR YOURSELF OR FOR OTHERS -

There are three main sequences to open the chakras. The first is to start with the first chakra at the base of the spine and systematically work upwards to the crown chakra. The second is to start first in the heart chakra. This will center the person that you are working with and keep them balanced as you continue. After the heart, open the navel point. This helps keep the attachment to the physical plane as well as starts to build astral plane connections. Then open the third eye to balance with the heart center. Next, work with the throat and then the crown center. After these top chakras are open, work with the second and first chakras. This latter sequence helps keep the person balanced evenly between the earth and sky polarities the entire time you work. It also avoids over-stimulating the bottom centers before they are balanced with the top ones. The third sequence is to open the top four chakras from the heart up to the crown. Then open the bottom three chakras if needed. These latter two methods are good to use if the lower centers are already fairly open or stimulated. If the person you are working with has any of the lower centers blocked, use the first method.

The following are exercises that will help you to open all of your chakra points:

EXERCISE FOR THE NAVEL CENTER

Sit with your spine straight, either cross-legged on the floor or on a chair with your feet flat on the floor. Rest your hands on the top of your thighs near your knees with your thumb touching your first finger on each hand. You can focus your eyes in either of these two ways: you can have your eyes closed and focused on your third eye. Or you can have them 9/10 closed and gaze at the tip of your nose. After you have centered yourself, say the following sound out loud continuously for at least three minutes. (You can later increase this to 7, 11, or 31 minutes.) The sound you use is HA RA. As you say this your lips should be slightly parted. Do not move your lips at all as you say the sounds. Use your tongue only. Your tongue will naturally touch the roof of your mouth as you say the "R" sound. (The HARA sound will slightly change as you do it to sound more like HA RUH with the "R" slighly trilled as if you are speaking Spanish.) As you say the "HA," contract your navel point back to your spine as if you are pushing the sound out. Release it on the RA sound.

This should be done in a rhythmic continuous manner. Let your voice go up and down in volume as you let the sound carry you. When you have reached the end of the exercise, inhale, hold slightly and exhale. Relax for a few minutes. If you like you can amplify this by wearing a yellow citrine or a clear quartz on your navel point. (Put it on a chain, cord or a belt buckle.) The tip should point upwards if it is single-terminated.) You also can put a clear crystal of like power and size in each palm as you keep your thumb touching your first finger.

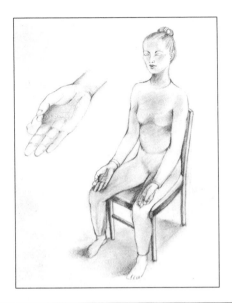

EXERCISE FOR THE HEART CENTER

Place your palms flat together so that your hands are touching completely from the palms through the fingertips, as if you are saying a prayer. Take your two hands and press them to your heart center, with a firm pressure. The position will become automatic. You won't have to think about it after a while. As you're sitting straight and relaxed, calm and centered, feel the pressure in your hands and focus on your heart. This exercise units the male/female or sun/moon energy in your body, which centers you. Breathe in and out of your heart with long deep breaths.

Then sing these words to yourself or out loud: RAH MAH. Sing these notes out of the center of your chest where your hands are placed. You can circle the RAH MAH sound out of the front of your chest over your head to the back of your chest. Then reverse the circle, sending the sound out from the back of your chest over your head to the front. Remember, the heart center is on both sides of the body, front and back. Vibrate your heart center with the sound. You can use any tone you like. RA refers to the sun. MA refers to the moon. The AH sound is the sound of the heart center. Do this exercise for 3 minutes, 7 minutes, 11 minutes, or 31 minutes. Then inhale, exhale, and relax. If you want to use your crystals to amplify this exercise, surround yourself with a circle or six-pointed star of clear crystals or emerald green crystals. You can place a crystal between your palms as you hold your hands in position. Wear a crystal necklace over your heart center. Use either a rose quartz or green or clear crystal right over the heart center where you place your hands.

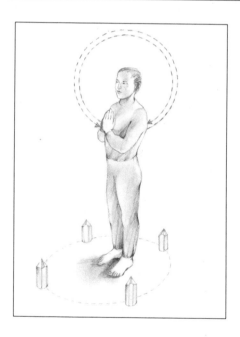

EXERCISES FOR THE THROAT CENTER

Here are two exercises to open the throat center. First, sit in a relaxed manner with your spine straight. Center yourself. Close your eyes and focus on the throat center. Next, begin to breathe with long, deep breaths. Have the breaths seem to go in and out of the throat chakra. Do this for 11 minutes. When you are through, sit for a few moments and ground yourself.

This second exercise you can combine with the first one or do it by itself. Sit with your spine straight. Center yourself and close your eyes. Maintain your focus on your throat center. Sing the sound HU (Pronounced as in WHO). Seem to vibrate your throat chakra with the sound. Hold the sound as loud as you are comfortably able to, then exhale before starting again. The rhythm should be continuous.

EXERCISES FOR THE THIRD EYE

To open the third eye, wear a crystal headband with a herkimer diamond crystal or amethyst crystal on an upward-pointing triangle, or an Isis moon on the cobra head of the kundalini serpent. You can sit and breathe in and out of the third eye center like you did in the heart opening exercise, vibrating that center with the breath. Or sing the sound EEE through the third eye center as you maintain your focus on it.

The following breath meditation opens the channel between the third eye and the heart. Sit with a straight spine and your eyes closed. Center yourself. Breathe in through your heart and exhale the air up and out of your third eye. Do this for at least three minutes. Then breathe in through the third eye and out through the heart on the exhale. Do this for the same amount of time that you did the prior direction. Hold a single-terminated or double-terminated crystal in each hand. The color can be clear, amethyst, or light gold. Allow the crystals to rest on your knees. You can open your hand in an ancient hand position called gyan mudra, with the thumb touching the first finger and the palms up. Rest the crystals in your palms. (See illustration L1.) The two crystals should be equally balanced with each other in terms of size and energy. If you like, you can create a ring of crystals around you or place crystals in the four directions around you.

EXERCISE FOR THE CROWN CENTER

Sit upright in a relaxed manner with your spine straight as in the prior exercise, L2. If you are sitting on a chair, place both feet flat on the ground. Rest your hands on the top of your legs near your knees. Touch your first finger to your thumb and have your palms facing up. Place a herkimer diamond, clear or amethyst crystal in the middle of each palm. Center yourself, close your eyes, and begin to breathe with long deep breaths through your nose. With your eyes closed, focus on your crown center at the top of your head. Now begin to vibrate the center with your breath as you seem to breathe in and out of it.

As you stimulate the crown chakra with your breath, visualize brilliant, golden light. Surround yourself with it, inside and out. Imagine that as you look up to your crown center the golden white light continues indefinitely. Continue this breathing/visualization process for at least three minutes. Now, as you continue to visualize the golden light begin to sing the sound OM. Imagine that this tone vibrates the top of your skull. Continue to repeat the sound "OM" in a steady rhythm. Let the sound guide you to establish the speed with which you want to repeat the OMs. Do this for the same amount of time that you did the earlier breath process, at least three minutes. You can do each section for 3, 7, 11 or 31 minutes.

KUNDALINI ENERGY

What is kundalini energy and how is it experienced? Kundalini energy has been described in many ways: as the life force of the universe, as Christ consciousness, as the supreme potential of man, as shakti or the feminine creative force of the universe and the nerve of the soul. People who have had the experience of the awakened kundalini generally seem to find it so wondrous that they are at a loss to find the words to completely express it. Kundalini energy often begins to rise in crystal work activating each chakra. As they open, their attributes can be utilized in your crystal work. Kundalini can be activated with quartz crystals enabling you to experience the accompanying consciousness expansion.

The kundalini is initially coiled three and one half times at the base of the spine. As it awakens it spirals upward, piercing and activating each chakra, then out through the crown center at the top of the head. When this happens, you feel as if you are blissfully merged in an ocean of pure, golden consciousness which is impossible to express in words. The process of kundalini rising is experienced in many ways. In fact, though there are certain generalities, each person's experience is unique. The length of time it takes to happen also differs for each person. Kundalini rising may be felt as heat, or liquid fire, especially along the spine or over the top of the head. It may be felt as pressure or tightness, especially near the various chakra areas. You may feel dizziness or shake. The process may be temporarily fatiguing to the physical body. (Thus, good health is a necessity.) The intuitions expand. Various states of higher (and different) consciousness begin to be perceived, experienced and intertwined with "normal" consciousness. Sometimes when this happens you may feel as if you are "going crazy" especially if you have no outside guidance. With time, though, you learn to integrate this into your daily life.

As the kundalini rises, different powers tend to develop. These can help you in your crystal work though they usually are not to be spoken of or used unless you are guided to by your "inner voice." Use them wisely. These powers tend to be those of clairvoyance clair-audience, psychic healing abilities, astral projection, freedom from disease, the ability to communicate truth, the ability to send kundalini energy to others, and other seemingly miraculous powers. One whose kundalini is awakened doesn't necessarily experience all of these powers, sometimes just one or two. This is because the kundalini vitalizes certain centers more strongly than others to allow you to help others in ways that are most appropriate for you. The truest test, in fact, of whether kundalini is awakening in you is that with this awakening comes a compelling or overpowering urge to help others. The fact that you are even interested in quartz crystal work shows some signs of the beginning of a kundalini awakening.

The point of awakening kundalini, however, is not to gather powers, as useful as they may be. The purpose of kundalini awakening is to be able to dwell in a state of the highest consciousness. The appearance of powers serve as signs of a raising consciousness and awakening kundalini energy. Dwelling on them alone without the proper focus on this highest development of your inner self is a mistake. It will lead you astray. Powers or metaphysical abilities are not the path itself -- they are merely the sign posts along the path. Treat them with lightness, wisdom, and joy.

How do you prepare for the awakening of the kundalini force? Your physical body must be healthy, especially having a strong nervous and glandular system. You also must have developed yourself sufficiently so that your focus is fixed firmly on your higher self and your natural attraction to help others. You must have developed your will to guide this energy and keep it focused on your higher centers. Again, the fact that you are interested in quartz crystal work usually shows some degree of interest in your higher self. Your kundalini energy will rise automatically as you become ready. Be patient.

What can happen if the kundalini is forced to awaken prematurely without the proper guidance and preparation? Some effects are purely physical. Its uncontrolled movement can produce intense pain and even physical injury as it bursts its way through any physical, emotional or mental blockage. The awakening of kundalini intensifies everything in your nature. Therefore, if it awakens prematurely it can intensify lower qualities such as pride and greed, instead of the higher qualities you would like awakened. These warnings are not designed to scare you in any way or to make you wary of the kundalini awakening. If aroused in its own proper time the results are wonderful beyond description. These warnings serve mainly to inform you that it is best to prepare yourself for the kundalini awakening at the proper time, taking its natural course in terms of your own evolution and purpose.

Your quartz crystal work will prepare you for and encourage the kundalini awakening. However, because you are not focusing specifically on the arousal of the kundalini, but instead, on the work you are doing, any arousal will be natural and gentle. The crystal work automatically helps you develop your higher self because if you do anything with the stones that is wrong or ill-intended it will automatically come back to you. You also automatically become attuned to your own "inner voice" through the crystal work. Your actions become strongly centered on this inner or truthful attunement because you feel so much more deeply content doing this work that becomes your way of life. This natural attunement will help to guide any opening of kundalini so that it becomes a source of growth and expansion instead of one of pain or fear.

EXERCISE TO STIMULATE ALL CHAKRAS AND PREPARE FOR KUNDALINI AROUSAL

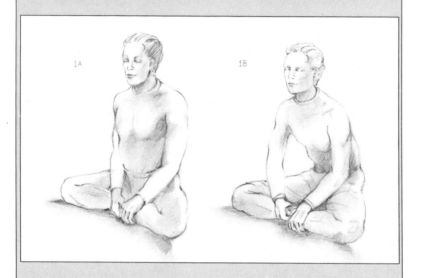

1. Sit cross-legged, in lotus, or half-lotus position. Spine straight, head forward. Grab the ankles with both hands, deeply inhale as you flex the spine forward and lift the chest up (1A). On the exhale, flex the spine backwards (1B). Keep the head level so it does not "flip-flop." Do not rock back and forth with your hips. Repeat this 26 times at the pace of one complete inhale and exhale every second or so. Feel the environment around you and inside your body. Relax.

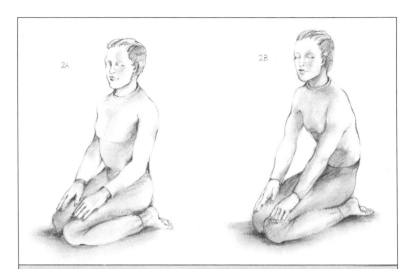

2. Next, sit on your heels. Place the hands flat on the thighs. Flex spine forward with the inhale (2A), backward with the exhale (2B). Think "Sat" on the inhale, "Nam" on the exhale, or "RA" on the inhale and the "MA" on the exhale. Repeat 26 times. Do this at the same pace as the exercise above. Rest about 30 seconds

3. Sitting cross-legged, grasp the shoulders with fingers in front, thumbs in back. Inhale and twist to the left, exhale and twist to the right. The breathing is long and deep, and at the same pace as the exercises above. Continue 13 times and inhale facing forward. Sit quietly about 15 to 30 seconds.

4. Next, hold your hands in this grip. (4A) Place the left palm facing out from the heart center with the thumb down. Place the palm of the right hand facing the chest. Bring the fingers to be against each other, together. Then curl the fingers of both hands so that the hands form a fist. You hold the hands opposite the heart center about 4" (4B). Form a slight pressure of seeming to try to tug the hands apart. You will feel a pull across the chest. Move the left elbow up to the left ear as you inhale

through the nose. Exhale while you bring that elbow down at the same time that you raise the right elbow up to your right ear (4C). As you do this, keep your hands centered at your heart. This resembles a see-saw motion. Do this in the same rhythm as the above exercises. Continue 13 times and inhale. Then exhale and pull up on the anus, sex organs and navel point briefly. Relax about 15-30 seconds.

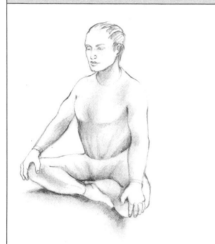 

5. Sitting cross-legged, grasp the knees firmly and, keeping the elbows straight, begin to flex the upper spine. Inhale forward, exhale back, as you did in the previous exercise. As you do this be sure not to bend the elbows. Keep the head facing straight forward. Repeat 26 times, rest 30 seconds.

6. Shrug both shoulders up with the inhale, down with the exhale. Do this for less than two minutes. Inhale and hold 15 seconds with shoulders pressed up. Relax the shoulders. Again, do this in the same rhythm as the prior exercises.

7. Roll the neck slowly to the right five times, then to the left five times. Inhale, pull the neck straight.

8. Lock the fingers in the same position as the fourth exercise. Hold them at the level of the throat. Pull the hands gently as in the fourth exercise (8A). Inhale and as you hold the inhale pull up with your anus, sex organs and navel. Hold for as long as you are comfortably able. This forms a lock so the energy you are generating doesn't "escape" from the bottom of the spine but is sent upward. Call this the "root of the spine" lock, (or root-lock for short). Exhale and again do the root lock as long as you can comfortably hold the breath out. Then raise your hands, still together over the top of your head. Do the same breathing process: inhale and comfortably retain your breath as you pull slightly on the hands. Then exhale and again do the root lock: pull on your hands as you hold your out breath. Repeat this cycle two more times.

9. Sit on heels with arms stretched over the head (9A). Interlock the fingers except for the two index fingers which point straight up (9B). Say "Sat" or "Ra" and pull the navel point in: say "Nam" or "Ma" and relax it. Continue at least three minutes. Then inhale and squeeze the energy from the base of the spine to the top of the skull as you pull the root lock.

10. Relax completely on your back for 15 minutes.

This series works systematically from the base of the spine to the top. All 26 vertebrae receive stimulation and all the chakras receive a burst of energy.

Each exercise that lists 26 repetitions can be done 52 or 108 times, as you become stronger. The rest periods are then extended from one to two minutes.

1. On every exercise, if you like, repeat the words Sat on the inhale and Nam on the exhale silently to yourself. Or, silently repeat the words RA (sun energy) on the inhale and MA (moon energy) on the exhale.

2. Between exercises maintain the state of awareness that you are in. Don't drop your focus. These exercises are in a particular

order, allowing a certain process to happen. In this process each exercise and resulting state of awareness build on each other to produce a certain state of mind, that of the highest awareness. This awareness not only helps you in your crystal work, it also creates happiness. How is this awareness produced with these exercises? The combination of position, breath and (inner) sound set up the conditions by which you are able to break through any blockages. This happens to the degree you remain willing and open, and surrender yourself to the process. Also, changes happen to the degree to which you maintain your focus.

    3. These exercises help you to break through blocks by arranging your body to set up certain energy pathways, placing your mind in a particular state of awareness. You will break through the blockages which keep you from fulfilling your highest potential and living in a state of contentment and inner fulfillment.

THE ASTRAL BODY AND ASTRAL TRAVEL

The astral body is somewhat contained within and extends beyond the etheric body. The more developed the astral body, the further its extension from the physical. This body has a rate of vibration that is higher than the physical and etheric bodies but is lower than the mental and causal bodies. The astral body is the vehicle of sensation and emotion. Whenever you express an emotion you are using your astral body whether you are conscious of it or not. Every feeling instantly affects this body and is reflected in it. Every thought that affects you personally is also reflected on this body. Your astral body not only responds to that which is directed from your physical body, but also to that which is directed from your mental body, both consciously and unconsciously.

Unlike your etheric body, you can consciously inhabit and use your astral body apart from your physical body. This happens automatically in your sleep. However, your ability to use your astral body apart from your physical body in sleep as well as in a waking state depends on your ability to be conscious in it. To be conscious in your astral body you need to be able to build a bridge or a connection between that body and your physical body. This etheric bridge allows you to operate in your astral body and remain conscious as you shift back to your physical body. Thus, you remember all your experiences and information and can communicate and make use of them in the physical. When the etheric bridge is built and you are able to remain conscious in your astral body there is no difference between the sleeping and waking states, or death and life. There is a continuous stream of consciousness in which you are always alive.

The astral plane, of which the astral body is a part, is an entire universe which occupies the same space as the physical universe. Because it vibrates at a much higher rate than the physical universe, normal physical senses are unable to apprehend it. Though the objects and events on this plane are shaped and formed by the imagination and thoughts of those on it, it has an independent existence apart from the mind. It operates according to its own astral laws just as the physical universe operates with its laws. It has lights, sounds, and colors which don't exist on the physical plane. Perception on this plane can differ from physical perception and can be quite confusing until you learn to use your astral senses. You are able to sense this plane as you raise your body's vibration and develop your intuitive, inner awareness. However, to actually be on this plane you need to develop consciousness of your astral body and free it from your physical body. With your astral body you travel to the astral plane.

Through crystal working you can develop sensitivity to emotions and then the astral plane itself. To then work with this plane you need only use your astral sensitivity rather than using your astral body. You do this by adding an emotional component to your work. You choose the appropriate emotions that will stimulate the emotional/astral body directly. This will then indirectly affect the physical body. As has been explained earlier, changes in the astral body will change the physical body because of the dynamic interrelationship on a vibrational level. Emotion can most easily be utilized when working with thought projection. When you project a visualization or thought, also project the state of emotion with which it seems to most closely resonate. For example, when you are healing, notice the emotional state that seems most closely associated with the illness. Then use your will to change that. If you change the underlying emotional state, you can heal the illness completely or at least help its healing.

The techniques that you can use to do this are many. However, in all of them you must use your intuitive voice or inner knowing to sense what the emotion is. Then see what it feels like in your body and change the feeling to a more appropriate one. Then project the altered emotion back to the person. All of this is, in essence, changing the vibration. (See the section on emotion for more detailed explanation.)

Though working with emotion can be quite effective, you can do even more by working directly on the astral plane with your astral body. You can then work directly with someone's astral body using your astral body. You can train yourself to become awakened to self-conscious activity on the astral plane using crystal gazing techniques and crystal dreamwork.

The following is a guided crystal gazing technique that will enable you to shift your consciousness from the physical to the astral plane. This is sometimes called "leaving your body." As you do this exercise you will be guided into your astral body. You will go through each stage of learning to use it. Finally you will be led to a meeting with a guide who will take you through and teach you about the astral plane itself. Each step of this process will give you more experience and information about operating in this realm, enabling you to do the astral work that you are guided to do.

CRYSTAL GAZING TECHNIQUE FOR ASTRAL CONSCIOUSNESS

(First read the Chapter on crystal gazing and practice the techniques.)

Part A

1. Sit with a straight spine in a position that does not require you to focus on supporting your body or in any way distracts you from a state of deep concentration. Gaze into a clear crystal which is placed before you. The crystal should be at least 3" tall by 2" wide, preferably larger. You can also use a crystal ball. Have lots of light illuminating the crystal. If you like, darken the room. If you want to amplify this gazing technique, wear a crystal over your third eye point, hold a crystal in each hand and/or surround yourself with a circle of crystals.

2. Close your eyes and center yourself.

3. Gaze with relaxed concentration into the crystal in front of you. Find a doorway or a place that visually interests you. Gaze at it more closely, noticing every detail. Continue doing this allowing the crystal to draw

you more and more into it. At some point your eyes will naturally want to close. Let them close and at the same time feel as though you are inside the crystal

4. Continue to maintain your focus as you are inside the crystal. Visualize yourself becoming luminous as you focus on the top of your head. Become more and more filled with light. As you are aware of this luminosity feel as if you are becoming lighter and lighter, so light that you begin to rise within the crystal.

5. Visualize yourself rising up to the top of the crystal and out of the tip.

6. See yourself in your mind's eye rising up to the top of the room. Look down beneath you and see your body beneath you. See the luminous cord that stretches between you and your body, connecting you.

7. Gaze around you at the objects in your room.

8. Now, using your will, begin to lower yourself back towards the crystal. When you are over the top of it, slowly lower yourself back through the tip until you are inside the crystal.

9. Settle yourself to the bottom of the crystal. Direct yourself to the space around you. Then find either the doorway in which you entered the crystal or another exit and leave through it.

10. See yourself return to where you are seated in the room. Remember the room around you. Next, feel the surface on which you are seated. Become aware of your breathing, then your entire body.

11. Ground yourself. Clear yourself, the room and the crystals.

Part B

When you can consistently and comfortably do the above process then add the following to it after step seven

1. When you can see your body beneath you, the etheric cord connecting your two bodies, and the objects

in your room, then see if you can "fly" your body about the room. Use your will to do this.

2. Practice moving your body from side to side and up and down.

3. Roll your body over back to front.

4. See your connecting cord stretch thinner and longer as you put more space between your bodies. See it thicken and shorten as you close their distance.

5. Return to your physical body by the same process as above.

Part C

When you can consistently and comfortably move your body about the room, you will be ready to do this next process. However, do not move on to the next process until you can, without hesitation or impediment, use your will to power yourself. If there is any doubt or fear experienced at any point continue to practice part A and B. There is no hurry. It is best to build a very firm foundation before you proceed. It is something like learning to swim. You first get used to the water and learn to move your body in shallow water. Only after you have mastered that do you learn the various swimming strokes in deep water. What would happen if you suddenly threw yourself into deep water, not knowing a thing about swimming because you wanted to be "advanced"? The same principle applies in learning to maneuver in the astral world. It is a large universe with its own laws and realities which must be understood. When you are ready, you may begin the next step in this process:

1. After following the steps in Part A until you are out of your body, move your body about the room (Part B). Then leave the room. As you experiment with leaving the room you will notice that you can either leave through the door or simply pass through the wall. Again, use your will to do this. You will learn that the barriers in this

physical universe are not in the astral universe. On the astral plane you can jump over mountans and pass through any object. The strength and focus of your will determines what is possible.

2. Now that you have left the room, leave the building you are in and travel around the neighborhood. Again experiment with powering your body. How do objects and people appear to you? You can pass through a person without being noticed unless they are particularly sensitive. In fact, usually animals will sense your presence before another person will! (Remember how all planes are contained within each other.)

3. Notice what is happening around you: sounds, objects, people and events. Notice them in detail and remember them. Intend or will strongly that you will remember them when you are back in your physical consciousness.

4. Now, notice the thin filament of etheric substance which connects this astral body with your physical one. See how it has stretched thinner than when you were closer to your physical body. You are always connected to your physical body with this cord.

Part D

1. After you have become thoroughly familiar with the use of your thought and will in your out-of-body experiences around your neighborhood, choose a place anywhere on this planet and transport yourself there. You will find that even though you are covering much distance, you can be there almost instantaneously.

2. Practice lowering your astral body to the ground and walking around in this location.

3. Open yourself to hear the thoughts and feel the emotions of the other beings in that location. Observe and listen and explore. In your astral body, with its astral senses, you will be particularly sensitive to that which is around you on subtle as well as more gross levels. Use this experience to learn.

> 4. Then use your will to transport yourself back to your physical body which rests in the center of the crystal. Lower yourself through the crystal tip into your body. Follow the rest of the procedure in Part A.

Part E

When you are comfortable and experienced doing part D then you are ready to move up into the astral planes. As you move into the astral planes it is strongly advisable at first to have a guide with you to thoroughly acquaint you with all the aspects of that plane. There are advanced beings on the astral (and other) planes that are specifically there to help those who come to these planes after they leave their physical bodies behind in death. They will help anyone who is ready. If you ask for guidance and you do not meet one of these beings you are not ready to consciously move and work on the astral plane in your astral body. If this is the case, continue to work on steps A through D. Realize that the judgment of your lack of readiness is not a personal judgment of you. In fact, you will slow your progress if you take it as a personal judgment. Your physical body and mental/emotional structure must be well prepared before you embark on the next step. Perhaps your motives for wanting to work on the astral plane should be examined and clarified. Is your desire to work on the astral plane simply in response to an egotistical desire? Or is it in response to a higher will which directs you? Have you completely offered yourself in service and love to a higher purpose than your own personal desires? Be honest with yourself. This honesty becomes the fire of purification that leads to your own development. Be patient. When the time is right a guide will appear for you.

There are other things that are helpful for you to know before embarking on astral travel. As you wait for the appearance of a higher master-being to guide you, many other lesser astral beings may appear to you offering their guidance as well as other information. Do not make the assumption that just because a being is without a physical body or appears on another plane, that they are a "higher being," more advanced than you, more conscious, or in some way special. There is nothing necessarily more conscious in not having a physical body. Remember, people leave their bodies every time they die! They may be drawn to you by the force of your desire and though they may offer help in various ways they actually might have nothing to offer you. Or they may be detrimental to you.

Remember also, that they still may have the all too human propensity to lie, play tricks on you, or try to impress you as people sometimes do on the physical plane. There is another type of being on the astral plane that actually is just an astral body whose inhabitant has already left. The body has "died" on the astral just as bodies "die" on the physical plane. Left alone, the body will just disintegrate. Sometimes, however, these bodies are kept "alive" by the lingering thoughts of its prior owner or by thoughts of those on the physical plane who think that they are in contact with a being. These thoughts keep the body in movement and form and give it a semblance of "aliveness." Any (astral) speech of this body is merely the reflection of the consciousness or subconsciousness of those who are consciously or unconsciously powering it with their thoughts. You may unconsciously draw one of these "shells" to you with the intensity of your thoughts or desires. The more you accord it beingness with your interaction, the more it seems to speak.

There are many other beings on this astral plane but the two types described are ones which you may attract in your crystal/astral work. How do you distinguish between a bona-fide master who can be trusted to be your guide on these higher planes and a lesser being? You distinguish the same way that you distinguish between teachers and guides on the physical plane. Listen to your inner, intuitive "voice" inside. This voice is the same that you listen for in your crystal work. It will seem to appear as a sensing inside you, specifically in your heart area. How does your heart feel? The feeling in your heart when you meet a master will be unmistakeable. It is an exultation that may move you to tears. You will feel uplifted and more yourself, rather than less yourself. You will feel completely accepted to your deepest core. This is true of teachers, masters and guides in the physical plane as well as the astral. Many may claim to be a guide for you. Listen to your inner voice or inner sense. Do not be led astray by your emotions, your thoughts and desires or your impatience. (This prior information also applies in crystal work if you channel information from disembodied beings.) You can avoid many misunderstandings and problems by just "channeling" information as you sense it revealed inside yourself. As many ancient texts on the subject say, "you are the source of all wisdom."

You are not at the mercy of any being or event on the astral plane that you may meet. You can use your will to leave any situation or banish any being. Maintain yourself in a state of calm, centered, self assuredness, so that your vision is clear and the use of your will is completely accessible to you. Do not lose yourself in fear or doubt. It is debilitating. What do you do if you fall prey to your own fears or doubts? Implement the same procedures that you would on the physical plane. Begin the breath with long deep breaths as you focus on your heart center. Mentally say the sound "RA" on your in-breath and "MA" on your outbreath. (All events and beings will wait as you do this.) This will center, align, and balance you. You will then gain perspective and be able to act appropriately.

The following is a procedure through which you can meet a master/guide for your crystal/astral plane work. Do this after you have become comfortable traveling in your astral body on this planet and before actually moving into the astral plane yourself.

Do this procedure after doing the steps in part A up to step (3) where you are inside the crystal:

> Continue to maintain your focus inside the crystal. Feel yourself becoming luminous as you begin to vibrate at a higher and higher rate to match that of the crystal. See the crystal becoming more and more filled with golden white light as it vibrates at an ever increasing rate. Feel your skin tingle with the sensation of the vibration inside and outside yourself. As your vibration quickens and you become more and more filled with the golden-white light, feel that all your chakras are open to their fullest, filling

you with their energy. When you feel completely light continue to sit or stand on the bottom of the crystal even though you may feel like floating to the top. Focus on your heart center and at the same time earnestly request the presence of a master who will guide you on the astral planes. Communicate your purpose for wanting to work on that plane. As you prayerfully request the presence of your guide, fill your heart center with all your love and send out a beam of pink or green light from it as if you are rolling out a welcoming path. See it extend through the walls of the crystal further and further until you no longer see its end. Wait in this open, receptive state until you see a figure in the distance walking toward you on your extended, welcoming path of light. See this figure continue toward you through the walls of your crystal until this being is with you inside of it. Feel as if you are both surrounded with the pink/green light of the open heart. Commune with the being who is with you.

Introduce yourself although you are already known, having had your progress followed and guided for quite some time. Ask who this being is and any questions that you have. Again, reiterate your purpose. Trust your intuitions and listen to your heart. Then be still and let yourself be spoken to or guided. Perhaps at this first meeting he or she will take you out of the crystal on the astral plane. Perhaps later. Don't worry, he or she will now guide you every step of the way until you are ready to operate on the astral plane yourself. (Later, when you travel yourself on the astral plane, he or she will be available for questions and guidance if you need it.) You will find that there are astral crystals if you should choose to use them. They may be offered to you or may just appear to you. Use them as you would on the physical plane. Feel free to ask your guide any questions as to their use. Once you have used these crystals on the astral plane you can also use them when you are back on the

> physical plane by just re-experiencing your connection with them and then using them. These are called etheric crystals. Just visualize them and direct them with your will. At some point your guide will bring you back into the quartz crystal. After establishing how to best contact your astral guide in the future, follow steps 9, 10 and 11 in Part A.

The entire creation is contained within you

DREAMS, ASTRAL PROJECTION, AND CRYSTALS

Another method that you can use to develop the ability to travel and work on the astral plane is to work with your dreams. You can use your crystals to do this. The ability to control your dreams eventually moves them into the state of astral projection. How does this happen?

The sleeping state not only replenishes the strength of your physical body, but also rests your astral body. On its own plane your astral body is almost incapable of tiring. However, on the physical plane it soon tires of its exertion brought about by the interaction with your physical body. Therefore, to give both bodies rest, the two separate during your sleeping state. Unless you are using your astral body consciously, it usually hovers in the air above your physical body. You are not asleep, only your physical body. You are then using your astral body whether or not you are conscious of it. With slightly more consciousness of your astral body it may separate from your physical body further, floating about in astral currents meeting other astral bodies and having both pleasant and unpleasant experiences. Because you are not able to bridge your consciousness from your astral body to your physical body without a break you will, in this case, only remember bits and snatches or only a vague caricature of what really happened. Also, you may continue to think in your astral body, but you are so wrapped up in these thoughts they wall you off from anything outside of you.

The impressions that you bring back with you to the waking state are primarily or exclusively those of your own thoughts. If you are even more conscious in your astral body you are able to move about freely in it, meeting and exchanging ideas with others who are also conscious in their astral bodies. This may include friends as well as teachers and guides. You can learn and experience things that are impossible to you in your physical body. These are remembered only as vague bits of intuition unless you are able to remain conscious as you pass from your astral to physical body upon waking.

Usually there is a period of darkness or a blank, unconscious period of time as your consciousness shifts from your astral to your physical body. During this unconscious time, the memories of your experiences are lost to you other than as vague or confused impressions because they are not brought through into your physical brain. To bring these memories through to your physical brain you must be able to remain completely conscious during the shift from consciousness in your astral body to your physical body. To do this you must have developed your astral body sufficiently. Also you must have opened your chakras so that they can all bring the astral forces through to the physical. You must have raised your body's overall rate of vibration. In addition, your pituitary or third eye center must be actively functioning in order to focus the astral vibrations. All of these above requirements are automatically fulfilled as you develop your sensitivity in your quartz crystal work. The preparations for effective crystal work are the same as those for the development of this bridge between the astral and physical consciousness as well as for all other conscious astral work.

What are the preparations? Your nervous system should be strengthened and purified to handle the increased amounts of energy that will course through your physical body. Your chakras should all be open and their energy available to you. Your mind must be able to be calm and focused and able to consciously influence matter. You must have control over your emotions so that you are able to control yourself on the astral plane. (This is the emotional plane.) You can see that if you have done the exercises in this book and have developed yourself enough to sensitively and accurately do quartz crystal work, you can do astral work if you feel so guided.

The next thing that needs to be done is to build an adequate etheric bridge between the astral and physical bodies. This bridge is a closely woven net-like structure of etheric matter that allows the vibrations from your astral and physical consciousness to pass through to each other. When this bridge is built you will have perfect continuity of consciousness between your astral and physical life. Death as you have thought about or imagined it no longer will exist.

The following is a method that you can use with your quartz cyrstals to first build the etheric bridge that will allow you to remember your dreams. Once the etheric bridge is built this method will allow you to gain control of your dreams, thereby shifting them and yourself into the astral realms.

DREAMWORK METHOD

This method is very effective but takes patience as it can extend over a period of months to years. Don't take shortcuts or the process will not work for you. Do each step until you are well versed in it. The perfection of each step is necessary to build the etheric bridge that connects your astral and physical consciousness. The best crystals to use for dreamwork and the astral planes are herkimer diamond crystals. These are a particular type of crystal found in the area around Herkimer, New York, thus their name. They are particularly bright with great clarity. They have terminations on both ends of them. If you don't have access to a herkimer diamond crystal, select a clear double-terminated crystal that is extremely bright. Crystals like the above that include rainbows are especially nice to use as the rainbow gives some suggestion of the astral colors. For best results, the crystal should be at least one-and-a-half to two inches in length. Of course, if any other size crystal intuitively seems to be the right one, use it. If you are selecting a crystal for this process, clear your mind, center yourself, and focus on the fact that you want to use it for dreamwork. As you maintain your clarity and focus, select the crystal. Before you use it, be sure that you clear it. In between uses, wrap it in white, gold, violet or light blue silk. If you do not have silk, wrap it in cotton. Store it on the altar or a special place. Do not show this crystal to others or have them handle it. You want to make sure that it vibrates purely in harmony with you.

Step I: Each night before sleeping, program the crystal so that you will remember your dreams in the morning. Put it under your pillow so that it lies under your head. As you drift off to sleep, maintain your focus. Make this the last thought that you have before going to sleep. Keep a journal next to your bed in which you write down every dream that you remember the first thing each morning. Do this until you have a continuous record of remembering your dreams. (At least three months of continuously remembering your dreams.) You can skip this step if you already record and remember all of your dreams. Do not clear your crystal in between nights. This way the vibrations of your programming will continue to build.

Step II: When you can consistently remember your dreams, select a dream that has consistently reappeared or that has had a consistent pattern. Before you sleep, program your crystal (as in Step I) that you will dream this dream again. Place this crystal under your pillow. Maintain your state of concentration as when you programmed the crystal and slowly lie down in bed. Visualize youself as becoming more and more filled with golden light. See yourself becoming so filled with this light that it spills outward from your skin and extends outward from your body. In your mind's eye see the light extended outward about six inches from and roughly in the shape of your body. Concentrate on this form of light. See it as being separate from your body. Do not drop your concentration from this form as you drift off to sleep. This is the start of learning to control your dreams. Continue this step until you can consistently dream this particular dream. For at least 40 days record your dream the first thing in the morning. This is helping to build the etheric bridge. (Anytime you do not dream this dream, clear your crystal.)

Step III: When you can continually do Step II, clear your crystal and program it for your travel to a place that you have not necessarily been in your dreams but have been to in your waking state. Again put your crystal under your pillow and concentrate on your luminous form. As you gaze at your luminous form, maintain your focus on the place that you would like to visit in your dreams. Make this your last thought as you drop off to sleep. Again, record your dreams the first thing upon waking. Clear your crystal if you dream any other dream. Always reprogram it whether you have cleared it or not. When you can consistently visit the intended location for at least 40 days in a row, move on to the next step. You will begin to develop the power to astral travel.

Step IV: Next, program your crystal for your visit to a location in your dreams that you have not been to on the physical plane. Use the same process to program the crystal and drop off to sleep as you did in Step II. When you wake, record your dreams. Continue this process until you can visit this location at least 40 days in a row.

When you have mastered this last step you now have the ability to be anywhere or visit anyone on the astral planes. The etheric bridge will have been built for you to be able to move freely between the astral and physical plane with no loss of conscious memory.

As mentioned earlier, you do not have to be able to transfer the memories of your astral life back to your physical brain to be active on this plane. During your crystal work if you feel the need conscious awareness of your actions, you can still do this. The technique is as follows:

1. Just before you sleep at night, program your crystal that you will help in some way on the astral plane while your physical body sleeps. Either visualize or focus on a specific task or way of helping or offer yourself to serve in any way that is needed. Use the same type of crystal as used in the prior exercise.

2. After you have programmed it, place the crystal under your pillow.

3. Continue to focus on your intention to help as you drift off to sleep. If your concentration was strong you will help as you intended on the astral plane even if you have no memory of it on waking. This process will eventually build the etheric bridge so that you will be conscious of your astral life.

4. Do not clear the crystal upon waking. Let the vibrations associated with your request build up stronger every time you use it. However, if you do feel a strong inclination to clear it at any time, do so. Wrap the crystal and store it in silk or cotton as was explained in the prior dreamwork method.

As with your other crystal work, do not use these astral skills selfishly or in a way that causes harm. Do not pry into others' lives unasked. It will come back to affect you adversely. Use your astral vision and intuitive powers to discriminate between useful help and useless interference. Before you change things on the astral or physical universe, regard all of the implications and effects of what you are to do. See if there is a higher purpose being served by any suffering and/or events as they are. Is it correct action to interfere? This type of careful, discriminating action leads to great wisdom.

Now that the astral plane has been discussed and techniques given to develop the ability to consciously act on this plane, it is useful as well as important to consider the following question: Why would you want to concern yourself with operating on the astral plane in your quartz crystal work? Once you advance beyond simple curiosity and start working more and more with the crystals, you develop the inclination to help people. This automatically happens because as you work with the stones you develop yourself more. As you develop further, your higher centers, including your heart center, open more. As they open, qualities of empathy, love and compassion become yours. You also develop the vision that allows you to see a way of living for yourself and others that need not involve the myriad forms of suffering that exist. You see how you can use your quartz crystals to help yourself and others, whether it be to relieve stress, heal, energize or any other uses. If you can work on the astral plane you can be of immense help both to those beings on this plane as well as on the physical plane.

Often you can work far more actively with greater accuracy on the astral plane than you can on the physical plane. In your astral body you have a much deeper power of comprehension of what must be done to help. More information is available to you. You can travel anywhere, meet any teacher, and receive any instruction that you can later use on the physical plane. Much help is needed on this plane and there are those who are only too willing to assist you in your astral development if you have a desire to help.

Another benefit of being able to operate on the astral plane is that you learn from it. Your sense of reality expands past any prior limits you may have placed on it. Likewise, your sense of who you are and what you are capable of expands until you can literally be without limits. You no longer are afraid of death because you have experienced that you exist beyond your physical body, mind and emotions. You can dwell in a state of quiet inner contentment and fulfillment.

Death is an illusion . . .
You are always here now

THE MENTAL BODY, VISUALIZATION, AND THOUGHT FORM PROJECTION WITH CRYSTALS

The mental body projects beyond the astral body and is also contained within the astral, etheric and physical body. This body is ceaselessly in motion and constantly changing although tending to remain roughly in an ovid shape. The mental body has to do with the manifestation of the self as mind or intellect. Whenever you use your intellect, memory, or visualization you are using your mental body. As with the astral body, you can become conscious in your mental body and use it apart from your physical body. There is also a separate plane of being, the mental plane, which is a universe unto itself with its own laws and appearances. In the mental plane your mind becomes your vehicle, not your mind working as though in your physical brain, but your mind working apart from any physical matter.

The mental plane is made up of mental vibrations which create their own images. These images look like objects and beings to you if you were to travel on that plane. The sensititivity and development of your physical brain determines to which of these mental vibrations it can respond. It is the response in your brain to these mental plane vibrations which are perceived as thoughts. Thoughts, then, do not originate in your brain, but come and go depending on its receptivity. You can experience this mechanism yourself if you center yourself, still your mind and focus on your thoughts. For a more detailed explanation of this, see the sections on mind.

Every time that you think, imagine or visualize, you set up a vibration in your mental body. This produces two results: (1) radiating waves or vibrations called thought waves, and (2) thought forms. You use both thought waves and thought forms in your crystal work when you utilize thought project or visualization.

VISUALIZATION

A visualization is a picture that you build in your mind's eye using a series of inter-linking thoughts. Each thought creates an aspect of the picture. The larger or more complex the picture, the more thoughts are needed. Your concentration and clarity of mind are necesssary to hold the various aspects of the picture together to finally create one entire vision. The use of your will underlies the creation of the thoughts as well as the ability to maintain your focus. The visualization you focus on creates an actual thought form which matches that which you see in your mind's eye. It also creates thought waves. The quality and strength of your vision and thought determine the exact nature of the thought waves and form.

To visualize, you need to use two abilities: The first is the ability to maintain the clarity and focus of your mind. Development of this ability is discussed in Chapter 8, Training the Mind. The second is the ability to use your thought to create the pictures themselves which you view internally. This ability comes naturally for some, and for others it comes with some difficulty. This is a crystal exercise that can be used to develop your ability to visualize if it is not already something that you easily do.

VISUALIZATION EXERCISE WITH QUARTZ CRYSTALS

Sit comfortably with your spine straight opposite a large quartz crystal or a quartz crystal ball. If you like, darken the room and put some form of light behind the crystal for illumination. Close your eyes, clear your mind, ground and center yourself. Next, open your eyes and gaze into the crystal or crystal ball for one minute. After a minute look away from the crystal and close your eyes. Try to maintain the picture of what the crystal looked like in your memory. Next recall what you saw. Say it aloud or dictate it into a tape recorder. This process will help you to maintain the picture of it in your memory. Do this three times. Next, gaze into the crystal or crystal ball for two minutes and again try remembering what it looked like as you speak of what you saw. Next, gaze for three minutes, noticing all the detail that you can, firmly impressing the picture of the crystal in your mind. Again, see what you can recall. You can eventually increase your time to ten minutes and see if you can recall every detail that you saw in that time. This process will help you to retain ever larger picures in your mind's eye. Try and refer back to your memory the sight of the crystal rather than a list of thoughts about it that you have categorized in your mind. If you want to help yourself retain the image of the crystal in your mind's eye, wear a crystal over your third eye point as you do this exercise. You can either hold it there with your hand, or wear a crystal headband. Wear the point up.

The methods of projecting a thought or visualization are the same as if you were projecting an emotion. Use your will or strong intention, your breath and your crystals. Methods utilizing visualization are given throughout this book. Visualization is not just wishful thinking or fanciful imagination. It has the potential to be an extremely potent tool.

As with work on the emotional plane, there are many ways you can use your quartz crystals to increase the effectiveness of thought transference or the projection of visualization. First, using your will to project the thought or image through a quart crystal amplifies the original vibration. This allows the projection to travel further. Also, the amplified thought projection or visualization is stronger so it will affect the recipient more intensely. The crystal can be used to direct the projection, allowing it to travel more accurately to its goal. The quartz crystal can also be used to help you focus so that the thought is more clear and the thought form more completely shaped. This allows for more effective transference, thus better results. Knowing the exact mechanism behind effective thought transference and visualization projection will suggest a myriad of other uses with your crystal.

There is an important point to remember in your crystal work as you deal with emotions and thoughts. Even though your focus may be to project an emotion, visualization, or thought out from you, it still affects you to some degree. Your bodies will vibrate in harmony with that which you think of or feel, and naturally you must think of and/or feel something before you project it.

Of course, the effect on you lasts only as long as you allow it to remain, until you change your body's vibration with a new thought, visualization or other technique. However, even to change that in you which you do not like entails being able to experience it. If you resist the experience of it you will not be able to change it. It is wise not to project anything that you would be unwilling to experience yourself.

The following are two ways to use quartz crystals with the visualization process. They will increase your ability to work on the mental plane. The first is a method by which you can learn to transfer and receive thoughts between two or more people. This process is sometimes called extra sensory perception. The second is a guided visualization process that will further sensitize you to quartz crystals, open your heart, allow you to communicate with someone on the subtle planes and develop your subtle body. This second visualization process will allow you to experience the effectiveness of using it as a tool to work on yourself.

EXERCISE ONE
Thought Transference with Quartz Crystals
(Extra Sensory Perception)
An exercise for two people

Before doing this exercise it is best if you have read and experienced the chapter on crystal gazing. You also need to be able to clear your mind and maintain a stong degree of concentration. If you would like to amplify this exercise, wear a clear or amethyst quartz crystal over your third eye point. You can also hold a crystal in each hand.

1. Sit opposite each other with a crystal or crystal ball set between you at least waist high. You might want to darken the room and illuminate the crystal to make it easier to maintain your focus. Center yourselves a few moments before you begin. Choose who is to send and who is to receive.

2. The receiver maintains himself or herself in an open, receptive, clear state of mind while gazing into the crystal or crystal ball. The sender focuses on one clear thought or picture in his or her mind which is then projected through the crystal into the receiver. The sender should do this only after letting the receiver know that he or she is ready to send.

3. The receiver, while gazing into the crystal, waits openly until a picture or thought comes to mind. Do not intellectually think about this, or "figure it out." Pay attention to the first thought, image, or impression that comes to mind. It might be very subtle sense, or a clear impression. Without applying any judgment to what is seen or sensed, the receiver then tells the sender what was received.

4. The sender then tells the receiver what was sent. Check your results. Do not try to judge who was right or wrong. This is not the point of the exercise and will only interfere with the development of the ability to send and receive. Judgment entails using your intellectual mind which interferes with that part of mind that you use to work with thought transference.

5. Do the above process three times. Then the receiver becomes the sender, then vice versa.

6. Continue doing this exercise until either of you feels like stopping. Once you feel tired, rest. Straining yourself will only deplete you, slowing down your progress. To push yourself in this way entails judging yourself. Again, judgment is counter-productive. Just notice your results.

EXERCISE TWO
Individual Visualization Experience
(From Crystal Path Cassette Tape)*

1. Use a clear single- or double-terminated quartz crystal or a crystal ball for this exercise. Sit in a comfortable, upright position. Keep your spine straight so that energy can move throughout your body without obstruction. Close your eyes and focus on your breath. Breathe in and out with long deep breaths, completely filling and emptying your lungs. With each exhale, relax your body and center yourself. Keep your eyes closed during the entire time that you visualize the entire following process.

2. When you feel relaxed and centered, pick up the quartz crystal in your left hand. Hold the crystal as you press into it and rub it. What does it feel like? What is its temperature? Notice every aspect of the way the crystal feels to you.

3. Now as you are handling this crystal, evision it growing larger, floating out of your hand and dangling right in front of you at eye level. In your mind's eye see the crystal backing away from you, growing larger and larger. Envision the quartz growing so that it becomes the only object in your awareness.

4. Visualize yourself moving closer to the front wall of this crystal. The closer you move, the less aware you are of any edge to it. There is just crystal.

5. Rub your hands on the surface of this crystal. How does it feel to you now?

6. As you rub your hands, begin to rub your body up against the quartz. How does it feel? As you delight in its surface, you seem to just melt into the crystal in

*Crystal Path, U-Music, © 1985 Uma and Ramana Das U-103
P.O. Box 31131, San Francisco, CA 94131

enjoyment, until you are floating inside of its shimmering golden glow. Notice a slight coolness inside you. You might feel a slight breeze. As you float freely in the crystal, you feel peaceful and content. Let yourself go into your imagination, experiencing the inside of the stone.

7. Now, in your mind's eye, notice that there is light about you. See it as clear, bright white golden light. This causes your body to vibrate at a faster and faster rate. See yourself become even more intensified. Feel bright, vibrating fresh tingling light. Rise toward that light. Keep rising until there is just intense, clear, crystal clear white light about you.

8. You are now vibrating at an extremely high rate. As you notice this vibration, begin to relax into it. Become one with it and relax. As you do this, you begin to feel a softening in your center. Visualize your center. This is your heart center opening. Visualize your heart as a green glow that is edged with pink. The green is soft and clear.

9. Next, visualize that from the softening in your center extends a green, glowing path of light edged in pink. This path extends from your heart further and further, past the walls of the crystal as far as you can see.

10. Imagine that you feel very open and relaxed. Into that openness, invite someone with whom you would like to communicate. Invite them into the green glow with you. Welcome them. Extend your invitation to any one of your choice, a guide, a loved one, or one with whom you no longer communicate and would like to. This is your chance to communicate all those words left unspoken that you would like to speak. In all love and openness, welcome that being into the green pink glow of your heart.

11. Now see the other person walk toward you on the green/pink path. As they come toward you, they too become edged in green glow. They also feel open, soft and trusting.

12. Welcome each other, becoming one pink/green glow of openness. Begin breathing in and out of your heart, your center.

13. Begin to communicate. Listen, ask any question and speak. Openly communicate. Trust. Love and let go. Let there be nothing unsaid or hidden from each other.

14. After you communicate again become aware that the green about you is vibrating so rapidly that it seems you are dancing among a myriad of tingling particles. Relax, let go and enjoy the feeling.

15. As you dance in the green, bid the person farewell you are communicating with.

16. Envision them also saying farewell and then begin to back away from you. As they back further and further away from the crystal, you become aware of more green and pink between you. See the green and pink extend as they become so far away that you no longer see them.

17. Be aware of how you feel them in your heart even as their form backs further away. You no longer see them, but you feel them in your heart.

18. Thank them for being with you. Know that they love you and you them in spite of what may have been said.

19. Now feel your heart as a soft, open vibrating space.

20. Next, feel the vibration all about you. As you look about in your mind's eye, you see shimmering gold, white light. Feel the coolness on your skin. It is like a pleasant, gentle tingling wind. As you vibrate in harmony with it, notice that you seem to float freely in the vibration.

21. Feel yourself floating. Next, you gently whirl down, feet first. As you float downward it becomes warmer until there is only a vibrating soft heat about you.

22. Visualize feeling your feet against a surface. As you become aware of your feet against a surface see that you are again surrounded by clear crystal.

23. As you visualize gazing more and more closely at this clear crystal around you, you suddenly notice that it is in front of you.

24. See the shimmering wall of crystal in front of you. Rub your hand on this shimmering surface. Rub your body and notice how it feels against your body. Feel the surface of the crystal.

25. And now it seems time to leave the crystal, so back away. As you back away see the crystal growing smaller. Continue to travel further away from the crystal until you see it small in the distance.

26. As you look at it in the distance, the crystal begins to rise and float towards you.

27. As the crystal comes toward you, visualize yourself extending your left hand to accept it.

28. Now become aware of and feel the actual crystal in your left hand. Notice the temperature, the hardness, softness, and the edges.

29. How do you feel now? Notice your state of mind.

30. Relax. As you relax become aware of the environment about you. How does it feel?

31. Next, become aware of the surface on which you are sitting. How does it feel?

32. Become aware of your breathing and then slowly open your eyes.

33. Before you arise from this process, re-center and ground yourself. Notice your state of mind and emotion.

This visualization is particularly potent because you have included your feelings. Therefore, you are not only utilizing the mental plane, but the astral plane as well. In creating certain thought waves and thought forms, you have also created certain emotional states of mind. These thoughts and emotions may only affect you temporarily or they may be long lasting. The durability of their effects depend on the intensity of your focused concentration. If your focus was strong enough you have actually communicated with the astral and/or mental body of the being that you were with during this process. You used your mental and/or astral body to communicate with their mental and/or astral body. Even if you have not directly communicated with them from subtle body to subtle body, the thought waves or forms corresponding to your communication were projected towards them. These thought forms and waves will have some effect on their subtle bodies depending on the degree of their openness. If they are open and your focus was strong, the communication may even reach their consciousness. Though your rational mind may be telling you that this was all imagination, you have actually affected yourself and the other in some way. For further corroboration notice the difference in how you felt before this process and how you feel afterwards.

RECEPTIVITY, PROTECTON AND PSYCHIC SHIELDS WITH QUARTZ CRYSTALS

How important is receptivity to the process of projecting and receiving thought forms and thought waves? Can you effectively project thought if the recipient is not open? Also, is it necessary to protect yourself against free-floating thought waves and forms that you may not like affecting you? Can they actually harm you in some way? And finally, can anyone at any time project whatever they may into you? Isn't this an invasion of your privacy?

These are questions that are usually asked when you realize the nature of thoughts and intentional and unintentional projection. The most important fact to remember when asking any of the above questions is that neither thought forms nor thought waves can affect you unless there are vibrations in any of your bodies capable of responding sympathetically to them. This is true whether you are open or not. Thus, you see the importance of self-development. As you empty yourself of any lower desires and focus more on higher qualities, your body automatically vibrates at a higher rate. Any projection of a lower nature with its corresponding slower vibrational rate can not penetrate this more rapid vibration. Not only can it not penetrate through your body's more rapid vibration, but it gets repelled back to the person who projected the lower thought. Thus the best protection against any undesirable thought wave or thought form is to develop those higher qualities of peace, love, and contentment, etc. (This can happen naturally during the course of your crystal work or you can do specific exercises to develop yourself, as given throughout this book.) This natural protection holds true whether the lower thought form is intentionally directed towards you or free-floating around you.

If you feel that you are not sufficiently developed or are in some way still vulnerable to unwanted thought projections there are other ways to protect yourself. The first, most obvious method of protection is to avoid those environments which are likely to be filled with the thought forms that you may wish to avoid. For example, if you do not want to be affected by violent thoughts, don't go to a violent movie or activity that either puts those thought forms into the immediate environment or encourages participants to have violent thoughts. Actively create an environment around you that is of a higher nature. Be sensitive to and use colors, sounds, crystals, pictures, plants, etc., to do this. If you are feeling particularly open or in a situation in which you will be open, you can build a psychic or subtle shield around you. (Many people prefer to do this when doing a meditation or contemplation.) This shield is built on the subtle planes and will keep any unwanted thought or emotion from you. Finally, if you are aware that an unwanted thought is being projected toward you (as well as an emotion), you can use your will to actively resist it. The stronger your concentration and strength of will, the stronger your resistance.

This brings up the final point: even though you may be projecting a thought wave or form that is higher in nature, the recipient must be open to it and not actively resisting your projection. Therefore, it is important in your crystal work with thought projection, as well as emotional projection, to prepare the person to receive it. If you are sending from a distance, let the person know when you intend to begin and finish. If the person is present with you, have them relax. Then explain what you intend to do and the results that you both can expect. Guide them to a receptive frame of mind. Of course, if you are working only with yourself (as with visualization), the process of creating the thought in the first place causes you to vibrate in harmony with it. It is possible to affect someone with your thought projection if they are daydreaming or not actively using their mind. Usually any thought projection that you would use in crystal work will find some degree of harmonious resonance with the recipient. However, a neutral state of mind in the recipient is not as conducive to a successful thought projection as an open, receptive state of mind.

The following is a technique that you can use to create a shield against undesirable thought and emotional projections. This shield is in actual existence on the subtle planes about you and will protect your subtle as well as physical bodies.

METHOD TO CREATE A SUBTLE OR PSYCHIC SHIELD

1. Sit or stand with your spine straight. Hold a quartz crystal in each hand. If you like, hold or wear one on your third eye point. This will enable you to more effectively visualize the shield. Close your eyes and focus on your third eye point.

2. Visualize bright, golden light surrounding your body and extending outward about a foot in all directions. Be sure to remember to surround yourself under your feet and over your head. This golden orb of light surrounding you is roughly the shape of an egg with you in the center.

3. Visualize that any undesirable influence is bounded off of the edge of your golden egg of light with fierce outbursts of flame or fire.

4. See yourself remaining calm and unaffected in the center. If you moved in any way you are still surrounded with this shield of light, even as you put your crystals aside.

5. If you like, carry one or all of the crystals with you to remind you of and reinforce the creation and power of the shield.

6. When you no longer feel as if you need the shield, have the orb of light around you disappear in the same manner in which you set it up.

7. Clear yourself, and your crystal tools.

SUSTAINING A THOUGHT FORM WITH CRYSTALS

A thought form will stay alive and continue to affect its recipient as long as there is some form of attention behind it. There are times in your crystal work that you may want to continue a particular process for a length of time, perhaps in healing work. It would be difficult if not impossible to personally attend to the process for that length of time. Nor can the other person involved be with you that long. The following is a technique that you can use with your crystal to maintain and project a thought form or visualization extending over distance and a particular length of time:

1. Obtain a picture of the person that you are going to work with. If you do not have a picture, draw a picture of them as best as you are able. (It doesn't matter how well drawn or accurate your picture is.) Place this picture on an altar or any special place that can be left undisturbed. Use a quartz crystal that you feel would be good to work with in this case. Clear the crystal before you begin.

2. Sit or stand comfortably, clear and center yourself. Now, focus on the other person as you visualize the thought form that you wish to project. See every detail and maintain a clear focus. While you do this, hold the crystal in one or both hands and gaze into it as if you are sending the projection through your eyes.

3. Next, use your breath to project the visualization or thought into the crystal. Inhale, and project on the exhale.

4. After you have done this, place the crystal on top

of the picture of the person. Use your will to suggest that the vibrations of the crystal which now represent your thought or visualization continually project to that person through the medium of the picture. This will happen continually whether you are there or not.

5. You can now clear yourself and go about your daily life. Occasionally focus on the crystal from where you are if you ever feel as if the transference is weakening. (Use your intuition.)

6. You might carry another smaller crystal with you in your pocket or other container to remind you to occasionally "tune in" to the larger crystal.

7. A good length of time to continue this is 30 days. However, you can continue it for any length of time that you feel is appropriate.

8. When you are through with this process, take the crystal off the picture and clear it. Clear yourself again and do what is appropriate with the picture.

CHANNELING

The purpose of channeling is to bring down information from more subtle planes into this physical plane for other people and for yourself. Your purpose is to serve as the means (channel) by which the information comes through. Remember, service is the keynote for this as with any of your crystal work. Do this process consciously and with the best of intention.

The process of allowing yourself to be a "channel" of information is a follows: First you center, ground and clear yourself using any of the various methods suggested earlier. Then you prayerfully request from your heart center that you be allowed to serve as the instrument through which "higher" wisdom will flow. Sit quietly and be aware of a vibration in your body and the environment around you. Then as you focus, wait for an increase of this basic vibration within and around you. You might feel this as an inner sense, shivers, heat in your body, or in many other ways. If you don't perceive any physical change, inner sense, or other changes in the first five minutes, then do not go on with the channeling. There may be some kind of blockage in your physical, mental, or emotional bodies that is preventing the free flow of information. Try to empty your mind, center yourself. Then stimulate your chakra points to open them. (Try using the Energy Series on page 76.) If you still would like to offer yourself as an instrument, do so. Then, again, wait for an answering response. If there is no response, don't channel this time.

When you finally feel the physical changes or inner sensing described earlier you are ready to begin. To do the channeling itself there are four basic things to remember and work with. First, focus on the topic or question that you want to receive information about. This can be as broad or specific as you want. Next, ground yourself so that you always stay aware in your body. Some people feel as if they must be "taken over" by another being who proceeds to use their body. That is one technique, but is unnecessary and not recommended. Because you and any etheric being are ultimately the same, another being in your body is creating a duality where none exists. If, on the other hand, you feel there is another being in your body and you want to rid yourself of it, just kick it out! Use your will to make that being go, and at the same time send love out to that being. The love will help you to be forceful and firm. The next thing to remember is to keep yourself clear of any personal motivations and attachments to the information coming through you. See pages 221-225 on attachments. (Who are you anyway?) Don't allow doubt, fear, or self pride to interfere. As in all crystal and metaphysical work, personal motivations pollute the purity of the message, or stop it entirely. *You are most effective as a channel to the degree that you have given up your personal attachments.*

One good way to gauge the depth of your personal motivations or attachments is to notice whether the attention of others is focused on you or on the message you are delivering. Are you feeling important, loved, rewarded, etc.? If you are completely focused on your higher self or inner guidance, you will not feel unloved, unrewarded, lonely, unimportant, etc. (Neither will you feel important.) You will just be you. Finally, don't judge the information which you receive as a channel. Just communicate it. Later you can check the veracity of your information.

What does it feel like to channel? What do you listen for? When you are channeling, you listen to your inner voice, or that source of truth inside of you. Some people will hear voices and sounds or see writing before their inner eyes. Most people get impressions or subtle sensings that have no particular form or shape. Sometimes feelings accompany these impressions and reveal further information. Just stay in a receptive mode with a clear mind and communicate what these subtle impressions are. Start with the first thing that comes to you, even if it seems tentative. As you continue, these impressions will become more rapid and clear to you.

The feeling you have when channeling is as if the information or impressions flow through you. Sometimes you might seem as if you are using your imagination. Then you tend to dismiss what you think you are "just imagining" as being invalid. Don't worry about that. As long as what you are channeling corresponds or resonates with your inner truth, you are speaking truth. All imagination is built on a subtle plane truth anyway. *Speak of anything that resounds in harmony with your own sense of inner truth.* That harmony with your own inner truth is your judge of the truth of what you sense. If you have any doubt of the truth of what you are about to communicate, don't communicate it. Your accuracy will increase to the degree to which you align with the sense of truth inside you.

For the channeling process to start, it is necessary to first clear anything that would be a distraction in the space in which you intend to channel. The environment should be harmonious. The telephone should be out of the room. You should arrange not to be interrupted and you should have plenty of time for the process. You will not be completely open if you think that you might be interrupted. You want the place you use to be a safe space in which you feel good. You might also smudge the room with sage, cedar or sandlewood, using the smoke to clear it, as we talked about in an earlier chapter. If you like, use any rituals you enjoy, or place special objects or pictures in the room to help amplify the vibrations. Arrange anything you need beforehand so you won't have to lose that train of thought coming through you in searching for something. Use paper, tape recorder, or have someone to listen to you, depending on what works best for you.

CHANNELING METHOD WITH CRYSTALS

Here are some ways in which you can use your crystals to help you in this process. Before you begin channeling you can hold onto a large crystal, sometimes called a generator crystal, placing your hands on the tip or on opposite sides of the crystal.

This will energize you. You can also use crystal gazing for this technique. Gaze into either a clear crystal or a crystal ball. You may want to use amethyst, but clear crystal seems to allow the most variation in what comes through. (Refer to the chapter on crystal gazing.) Allow yourself to go into the crystal and allow information to come through from the tip of the crystal down toward you and into you. When inside the crystal, you can see information written on a screen in front of you. Another method is to bring a teacher into the crystal space with you. If you like, you can allow yourself to be in a large crystal room, filled with many other people like yourself who are asking questions. This crystal room is lined top to bottom with cubicles, some with people, some empty. You step into a cubicle, sit down, and find that this cubicle has a tremendous force inside it. As you meditate and open yourself to it, you find your answers. You can also enter the crystal and eventually see yourself going down a long hall with books lining the walls on both sides of you. The hall feels cool, fresh and clear. As you walk down the hall, the right book just seems to spring into your hand. There are endless methods to use inside the crystal. Be imaginative. Be creative. Set up something that will work for you. Finally, surround yourself with a crystal force-field created by a specific geometric shape that will amplify whatever you do. Hold a crystal in one or both hands as you work. Wear crystals on your heart, throat, and/or third eye. If you like, keep them wrapped and use them only when doing this work.

At some point, it will feel natural to stop channeling: don't try to grope around for more to say. Just stop. If you don't you will find it harder to channel truthfully. You can become quite drained or tired. When you are through, bring yourself out of your crystal if you were using that method. Ground and center yourself. Then clear the space you are in, others around you, and your tools. After channeling you usually find yourself filled with contentment and a very good feeling. Sometimes you might be physically, emotionally, or mentally tired. If that is the case, take care of yourself. Work on your development more to avoid depleting yourself and to be able to channel more easily.

DISCOVERING WHAT IS STORED IN A QUARTZ CRYSTAL

When you first come into contact with a quartz crystal you generally want to clear out any prior vibrations which it contains. It may contain certain influences that you would not like to affect you. However, many times a crystal has information stored in it that would be useful for you. Crystals exist that have been deliberately programmed with information for those who discover them at a later date. Some of this programming is recent, and some dates back to ancient civilizations. You can deliberately pass an image or information on to someone to recover at a later date. In these cases you want to be able to discover what is stored in the crystal rather than automatically clear it. The following is a technique that allows you to discover what is stored in quartz crystals:

1.   Touch both of your hands to the crystal. If you like, pull your hands away slightly while still maintaining a subtle connection with the stone.   Either way is appropriate.

2.   Close your eyes and focus on your third eye point. This point is located between your eyebrows roughly in the center of you forehead.

3.   As your hands are in contact with the crystal, begin to breathe in and out of your nose with long, deep breaths. On each exhale, relax more deeply.

4.   Let all thought which enters your mind go until you are just focused on the crystal.

5.   Feel as if you are vibrating in harmony with the crystal.

6.   As you vibrate in harmony with the crystal, have the strong intention that you will learn what is stored in it. Maintain that intention during this entire process.

7.   You will receive the information that is stored in the crystal in two ways:

> A. Your third eye point may start to flicker with light and you will see in it images that represent what is stored in the crystal.
> B. You will receive impressions of what is stored. They may be quite strong or very faint. Do not try to use your intellectual mind or you will no longer receive these impressions. Just remain in an open, receptive state.

8. When you are through, clear the stone if you want to do so. Clear yourself and your environment. (See "smudging" method.)

If you would like to deliberately pass on images or messages to another, you first program the crystal as explained in page 20. Then the other person uses the above process to receive the information. This is a good way for parents to pass on certain experience and messages to their children. A similar situation would exist between a teacher and his or her students.

CRYSTAL GAZING AND CRYSTAL BALL READING

Crystal gazing and crystal ball reading refer to the process of using the quartz crystal as a focusing device or as a starting point through which you develop and maintain an altered state of awareness in order to access certain information. This altered state of awareness, sometimes called a trance state, can allow you to delve deeply into the stored material of your subconscious to answer specific or general questions or to receive specific information. This trance state can also sensitize you to certain etheric vibrations so that you can "see" the future or past. It can be used as a method to begin to view the astral plane and travel on it. (See pages the pages on astral travel.) As a by-product of your work with crystal gazing or crystal ball reading you tend to develop your powers of concentration, will, visualization and telepathy.

There are many other mediums by which you can gaze or read. For example, these include gazing in water, black ink, tea leaves, magic mirrors or glass balls. However, the quartz crystal amplifies the etheric vibrations that you need to be aware of in some types of reading. The properties of quartz crystal help you to focus your attention more easily. The crystal energizes you, helping you maintain your focus long enough to get a clear reading. The quartz crystal also raises your vibration so that your trance state is deepened. Any ball shape creates a round field of vibration that tends to draw you into it as you focus on the ball. A quartz crystal ball only amplifies, quickens and deepens this process.

As with other crystal work, to be most effective it is necessary to develop your will power as well as clarity and calmness of mind. A harried, unfocused mind will not be able to see beyond the screen of thoughts that mask perception of the more subtle impressions. A weak and untrained will won't allow you the ability to concentrate or maintain a focus for any length of time. To be effective in crystal ball reading and crystal gazing you must also be able to to detach yourself from your everyday concerns.

To do this, develop a sense of yourself separate from that which you do, think or feel. If you do the excercises given in this book you will develop this sense.

To help you in this work, start to develop your ability to be aware of sensations or thoughts which may have come through the astral or mental plane. This ability begins to develop naturally as you crystal gaze or read crystal balls. Noticing incidents of these experiences during your everyday life hastens this process. Try to become aware of precognition and intuition in your daily life instead of automatically dismissing them. Finally, increase your ability to visualize or view mental pictures so that you can more easily see into the crystal. Exercises to increase visualization abilities are given. The methods for gazing into a crystal and a crystal ball are the same. When you gaze into a crystal ball, however, it is easier to narrow your focus to one point, thus deepening your trance state. As mentioned earlier, a crystal ball creates a round energy field with a whirlpool effect that tends to pull you inward towards its center. Because crystal gazing methods have been given throughout the book the focus here will be mainly upon crystal ball reading.

What type of crystal balls are best to work with? Some people prefer to work with crystal balls that are absolutely clear, feeling that anything else will distract them. Others prefer balls with formations in them called veils and inclusions. They feel it is easier to find an area that interests them and seems to pull their attention into the ball. This is called a doorway. Rainbows add value and color dimension to your crystal ball reading. They also replicate the colors of the astral plane to some extent and are used to help draw you to that plane. When choosing your crystal ball, select the one that you seem to be most attracted to or seems to draw you towards it.

With crystal ball gazing you need to look into it with relaxed concentration. Don't strain, but be very focused. If your mind wanders, gently bring it back. Gaze into the crystal ball with diffused or unfocused vision. You can best learn to use diffused vision in the following manner: Look straight ahead as you pull the outer corners of your eyes slightly toward your ears. Hold them that way and notice how you are seeing. Let go of your eyes and maintain that vision as you look into the crystal. With practice you can shift to this way of seeing automatically when you gaze into the crystal ball. Also, as your concentration on a point in the crystal ball deepens, you eventually pass into an altered state of consciousness in which you are aware of nothing but the point in the crystal where you gaze. Your vision automatically becomes diffused.

There are different methods of seeing into a crystal ball when you are in an altered state of consciousness. One way is to actually see visions. Usually in this method of seeing, the ball will seem to get greyish or cloudy. Next, a hallway will seem to open or curtains may part. Then you see your visions. Another method, similar to the first, uses the ball to become the starting point which opens up into an astral tube. This tube leads to the astral plane much like a telescope. You look into the tube and at the end you see your vision. In your trance state you can will yourself to go through the tube and actually enter your vision, viewing it in its entirety. Then you maintain your consciousness as you pass back out of the tube and out of your trance state so you don't forget what you saw. If you look at your vision through the astral tube rather than entering into it, you will see just what appears at the end of the tube. You will not see the other aspects which are behind or around it. So you must learn to aim the tube in all different directions. Use your will to do this. If you use the technique of traveling down the tube and entering your vision, you obviously don't have to aim the tube like this.

Another method is to gaze into the crystal ball and receive impressions or mental pictures. You receive these as a strong sense inside you. Don't judge them at first. Just notice something, however slight, and report it to another person, into a tape recorder or mentally to yourself. Don't automatically dismiss anything as being unimportant. As you begin to report your impressions they will feel stronger and seem to flow through you.

As with other crystal work, the more you pay attention to your intuition and work with it, the more it will increase in strength. As you work with the crystal ball you are building a psychic muscle system. Like any muscle system, the more you exercise and use it, the stronger and more trustworthy it becomes. The more you are able to report the mental pictures you see and the intuitive impressions you receive, the more accurate you will be.

As with all crystal work, check your results. See how accurate you are without judging yourself. Keep a notebook so you can see your improvement. As you become more accurate your confidence will improve, creating an even greater facility for successful crystal ball reading.

CRYSTAL BALL READING TECHNIQUE (1)

1. Darken the room around you. Put the crystal ball on a dark surface that will not reflect any light that will distract you. Have a candle or other form of light shining into and illuminating the ball. The light source can be reflecting up inside the crystal ball stand or behind the crystal. It can also be shining from behind you into the crystal ball.

2. Sit comfortably with your spine straight. Keeping your spine straight enables you to channel all the energy that comes through your body when you do your readings. Ground yourself and calm your mind.

3. Clear the crystal ball before using it. Use the smudging method discussed earlier. Then charge it using either of these two methods:

A. Hold the ball and rub it with both your hands. Focus on the ball as you inhale deeply through your nose. On your inhale feel as if you are inhaling the vitality of the crystal ball and mixing it inside your body with your vitality. Next exhale completely through your mouth into the ball, sending vital force from your body to mix with the vitality of the crystal. Visualize and feel this vital force being exchanged and joined.

Continue until you feel that you and the crystal ball are vibrating in harmony and that the ball is vibrant and "alive."

B. Hold the crystal ball and rub it with both of your hands. Then set the ball back on its stand while still maintaining your focus. Next, point a single-terminated quartz crystal or a crystal wand towards the ball at a distance where you most comfortably feel the two connect. When you feel or sense the subtle connection between the crystal and the ball, begin to circle the single-terminated crystal in a clockwise direction. Maintain the connection between the crystal and the ball. Soon you will feel the crystal ball pulsate. When you feel it pulsating strongly you can lay your single-terminated crystal or crystal wand down and begin the next step.

4. Diffuse your vision.

5. Find a spot in the ball that interests you. Bring all of your attention to that spot. Now look at it in more detail. Allow your intellectual mind to relax. Leave yourself open to any impressions. Don't have any preconceptions. Have no judgments. Remain relaxed, open and focused.

6. Now look at that area in still more detail. As you look at this area even more closely you may find yourself tensing. If so, take a deep breath and relax as you let it out. Continue this until you release your tension.

7. Continue to gaze into the particular area, seeing more and more detail. As you look at this minute detail the crystal ball seems to become larger. The more you lose your sense of yourself and are only conscious of the detail in the crystal space, the larger the ball seems to become. As you are looking, feel yourself opening as if you have no edges or boundaries. Allow yourself to lose your awareness of yourself and just be aware of the crystal ball.

8. Your eyes may or may not close at this point. If they feel like closing, let them close and continue to feel as if the crystal is around you. If your eyes remain open, still feel as if you are inside the crystal. Whether your eyes are open or closed a point is reached where there is no difference between you and the crystal. There is only crystal and just what you are seeing. You are now in an altered state of consciousness or trance state.

9. You may at this point find yourself feeling compelled to do something. This may be to breathe in certain ways, move energy in a particular pattern in your body, or to work with the crystal in a different manner. Trust yourself and let this compulsion guide you. Your body may almost involuntarily assume different positions. Let it. You may find yourself making sounds. Feel free to allow this to happen. At the same time there is no necessity to force it to happen. Successful crystal ball reading doesn't depend on any of these occurrences.

10. Now either ask specific questions in your mind or use your will to guide you to specific locations that you mentally view. If you haven't a specific question, mentally refer to your more general purpose in crystal reading at the time. Then let impressions come and go. Don't hang onto them but communicate them to another or out loud into a tape recorder. Let the impressions flow through you. Continue to communicate what you see until you have a feeling of being finished.

11. When you are through, remain with the crystal for a few minutes. Release any tension. You will feel a sensation of floating inside the crystal ball. Varied impressions may come and go. Just watch them pass. In this space are all possibilities, all assistance, all knowledge, and all wisdom.

12. Now, after enjoying this space for a while, slowly begin to back out of the ball. Start to feel edges around yourself while you continue to gaze. Have a feeling of backing up as you become more aware of the edges around the ball. If you followed a certain route into

the ball, retrace it. Continue until you see the ball in front of you. Become aware of the surface on which you sit. Be aware of your breathing. If your eyes are open, let go of the diffused vision and bring it back to normal. If your eyes are closed, slowly open them. Now stretch and shake yourself a little until you feel completely in your everyday consciousness.

13. Ground yourself. Next, clear your crystal ball, the other single-terminated crystal or crystal wand, yourself, and the room you are in.

14. If you like, cover your ball with cloth or put it in a pouch. Store it in a special place.

When you are gazing at a crystal ball and relaying information about what you see, you are actually a tool in the process. Your mouth automatically speaks what you are seeing without any personal involvement. Your awareness is not centered on what you are saying. When you speak there is a tendency to revert to the intellectual mind. If you do that, you will pull yourself out of your trance state. If you continue to speak while not in your trance state, your information will not be accurate. To keep that from happening don't judge what you are saying. If your mind wanders or you start listening to what you are saying, gently pull your mind back to your vision in the ball. You can tell when you revert to the intellectual mind because all of a sudden you will lose the vision. If you lose the vision entirely, repeat the process of entering into the crystal ball. Begin to see and communicate whatever comes to view whether or not it has anything to do with what you were speaking of before.

When you first begin this work, it can be difficult to sort through all of the impressions that you start receiving to focus only on those which apply to your purpose. If you start becoming confused with too many impressions that seem to be irrelevant, concentrate more on your purpose when reading the ball, or the particular questions to be answered. With strong focus, your impressions or visions will more closely reflect only those in which you are interested. Until then, choose only to communicate that which intuitively seems relevant.

The prior crystal ball reading method was one in which you receive and communicate intuitive impressions through the use of the ball. The next method will first explain the process of seeing an actual vision in the crystal ball and second, how to use the astral tube to see and enter your vision. The astral tube method can also be used as a device to enter and participate on the astral planes. The crystal ball serves as a doorway through which you can go back and forth between the astral and physical planes via the astral tube.

CRYSTAL BALL READING TECHNIQUE (2)

Again follow steps 1 through 6 as given in the first crystal ball reading technique. Then advance to the following steps:

7. At this point the ball may seem to get grey or cloudy.

8. As you continue to concentrate and gaze into the greyness or cloudiness you will see in front of you a clear, circular opening in the greyness. At the end of the opening is a vision which contains the information for you. Record this information. However, if you would like to know more, use your will to travel toward this opening. Go towards the opening.

9. As you go toward the opening you see that it is like a long hallway or tube that extends into the distance.

10. Use your will to enter into the hallway or tube. Follow to its end and exit into your vision.

11. At this point you will be conscious and able to operate on one of the etheric planes, probably the astral. Use your will to travel about and otherwise act in this environment as is described in the chapter about the astral plane.

12. Verbally speak about that which you see. If your focus has continually been concentrated on your purpose or question, what you see will reflect that.

13. It is preferable to have a tape recorder or another person to listen and record what you say unless you are well versed in maintaining a continual consciousness as you shift between planes. Until a continual consciousness is achieved, it is common to have a period of "blackout" or loss of memory when you shift from the astral back to the physical plane. (This is explained in more detail in the chapter on dreamwork.)

14. When you are satisfied that you have seen or done enough, mentally recall the opening of the hallway or tube through which you entered into this plane. Use your will to be back at this opening. Exit into this opening and again travel down the hallway or tube that stretches out in front of you.

15. Mentally see in front of you that the end of this tube or hallway opens into the greyness or cloudiness that you encountered earlier. Exit into it.

16. Use your will to travel through the grey cloudiness as it becomes progressively lighter and lighter until it is the clear crystal again.

17. Begin to back out of the clear crystal by the same path with which you entered. Continue until you see the crystal ball in front of you.

18. Refocus your eyes, returning to your normal vision. Become aware of your breathing, then the surface upon which you sit. Gaze around you at the room and the people and/or objects in it to re-orient yourself. Stretch and shake your hands, feet, and the rest of your body until you feel completely returned to normal consciousness.

19. Ground yourself.

20. Clear your crystal ball and the other crystal or wand that you used to charge the crystal ball. Next clear yourself and the environment around you.

21. Cover your ball with a cloth or pouch if you like and and store it in a special place.

To do the above techniques, begin by ball-reading for only ten minutes. When you find yourself able to easily maintain your concentration for this length of time without getting tired or otherwise straining yourself, increase it in five-minute increments. There is no set time limit in a crystal ball reading. The time is set by the limits of your strength and ability to focus.

Don't worry if you don't see anything in the crystal or don't receive any impressions, etc. Just patiently gaze into the ball for the time limit you've set. Some people see into the ball the first time, others take weeks or months. In some very rare cases, it takes years to read a crystal ball. Be patient. The rapidity with which you can start receiving impressions or visions depends on the degree to which you are developed in those necessary qualities. All crystal and other metaphysical work naturally develops those qualities that enable you to eventually read a crystal ball, and allow the method that is most suited to you to emerge. The method will find you.

To increase your effectiveness in reading the crystal ball you can place a crystal over your third eye point. You can hold it there although that is often distracting. Instead it is recommended to wear a crystal handband that holds the stone over your third eye point. The crystal will help stimulate and open this chakra, thereby increasing your ability to "see." You can also hold a crystal in each hand during the time you are crystal ball reading. This will raise your body's vibration so that you have more energy to maintain your concentration. Also the higher vibration will open and make accessible to you the consciousness in higher centers in your body without which you cannot "see" into the crystal ball. Always clear these crystal tools before and after each time you use them.

Finally, these three techniques are only meant to serve as a general outline or path for you to follow. Everyone has their own unique experience when they read a crystal ball even though they tend to follow one of these three explained above. Let yourself remain open to the unfoldment of your own unique variation of the crystal ball reading technique.

SEEING THE PAST AND FUTURE IN CRYSTAL BALL GAZING

It is said that you can see the future or the past when reading a crystal ball. How is this possible? All objects, thoughts, emotions, events, etc., which appear on the physical plane first manifest as vibrations on 'the subtle planes (i.e., causal, mental and/or astral plane). (This process is explained in detail in the extended body section of this book.) The vibration in the subtle planes can be seen as the cause which sets into motion the events which are to follow on the physical plane. When you are seeing into the future you are actually sensing or seeing that subtle plane vibration and interpreting the physical plane event which is to follow from it. You will be accurate to the extent that other subtle causes do not intervene. Next, anything that has occurred on the physical plane affects the vibrations on the subtle planes, leaving its impression. Eventually everything is "recorded" in the form of vibration in the akashic plane, or universal ether with which all space is filled. When you are seeing into the past you are seeing this latter vibrational pattern. This is sometimes referred to as viewing the akashic records. In any event, when you are seeing the past or the future you are just sensitizing yourself to and interpreting the vibrational or psychic "shadows" which precede and follow physical plane events.

PROBLEM SOLVING WITH CRYSTALS

Many times in your crystal work and in your daily life you encounter situations or questions that involve doubt, uncertainty or other kinds of difficulty. This is typically termed a "problem." Furthermore, the more you have attachments with regard to the situation on which the problem is based, the more your difficulty with it grows. The more attachments that are involved with a problem, the more emotions and conflicting thoughts seem to war with each other. Soon you find it difficult, if not impossible, to create the kind of clarity within yourself that will allow you to see the answer to the problem. To solve any problem you first need to see exactly why certain emotions and thoughts are involved and what purpose they are serving. Then you need to be able to extricate yourself from your personal involvement with the problem and refer to your higher guidance for an answer. When you are able to do this, a problem becomes less of an emotional or otherwise traumatic event. It becomes more like a puzzle in which you have only to fit the pieces together correctly to solve it.

Problems should become events or situations that come and go, requiring you to discover the most appropriate response. It is not that you no longer will involve your thoughts and emotions. Rather, you will no longer be the servant to them. When you are no longer their servant you are quickly able to clear up any doubt, uncertainty or other difficulty. You are able to clearly hear the wisdom which lies within you and be clear about what to do.

The following is a procedure that you can do with your quartz crystals to break through emotions and thoughts to a place of clarity that will allow you to see the answer to your problem.

GETTING RID OF PROBLEMS WITH QUARTZ CRYSTALS

1. Sit in a quiet space where you will be undisturbed. Have one quartz crystal with you. Use a clear crystal. If you are feeling emotionally overwrought or "hyper" use a light smoky crystal that will tend to ground and calm you. The crystal that you use should not be opaque, but clear -- one that you can see into.

2. Center and ground yourself. Hold the crystal in front of you with both hands while looking into it. (Do not diffuse your vision as in crystal gazing techniques.)

3. Talk your problem into the crystal. As you do this, the crystal will absorb all of the associated vibrations.

4. Next, clear the vibrations out of the crystal. As you clear the crystal, imagine the vibrations entering the earth where they will be transmuted. As you do this, feel the corresponding lift of your own vibrations. Your body might even feel lighter.

5. Clear yourself and the space you were in.

3

Tools on the Crystal Path

SOUND AND QUARTZ CRYSTALS

THROUGHOUT THIS BOOK emphasis has been placed on the use of sound along with various exercises or specific techniques of crystal working. What are the properties of sound that make it such a potent tool to use with crystals? As has been explained in this book, every object and person in this physical universe vibrates at a particular rate. That vibration is represented by a sound, only part of which can be heard by the human ear. A set of vibrations produced by a physical object or person has a corresponding set of sounds. Sound is, then, audible vibration. Thus, our physical universe has a corresponding universe contained within it which is entirely audible and inaudible sound. Furthermore, as explained earlier in this book, every subtle plane is made up of vibrations. These vibrations also have their corresponding sounds. So, as with the physical universe, every subtle plane contains or is represented by its corresponding universe of sound. There exists an entire universe entirely made up of sound.

Not only does vibration have its corresponding sound, but that sound creates a corresponding vibration in other planes which create other sounds. As has been shown earlier, a form created in the physical universe creates a vibrational pattern which affects the vibrations of each successive higher plane and so creates corresponding forms on those planes. Likewise, each form on a higher plane creates a vibration pattern which affects the vibrations in this physical universe to eventually create form. A sound on a subtle plane has corresponding vibrations on that plane which eventually manifests as forms on the physical plane. Likewise, a sound created on the physical plane affects the vibrations in the subtle planes to create form. So, if viewed from plane to plane, sound creates form and form creates sound. You can see then that sound, like quartz crystal, can act as a bridge between form and formlessness. Also, as with quartz crystal work, sound can be consciously manipulated to make physical changes. It also can be used in a conscious manner to link you to the astral, mental and other higher planes. You can see how it can be particularly effective to combine the use of sound with the use of crystal.

Before you can consciously work with sound you have to be able to hear it. Next, you have to develop an awareness of the attributes of each sound. The universe of sound is largely inaudible to the physical ear. It is primarily subtle sound with a small part manifesting as physical sound. How can you work with sound when you can only hear a small part of it? How do you hear subtle sound? Though the sound you are used to hearing is the physical sound of the environment around you, every physical sound has its subtle counterpart. So to work with subtle sound you first must become familiar with and work with physical sounds. As you work with physical sound in certain ways you can gain an understanding of and begin to sense subtle sound. You can eventually train yourself to hear the subtle sound which exists inside of you and all around you. These subtle sounds which fill all of space are sometimes referred to as the sound current. Just as you can develop the "inner eye" to see or sense that which is on the subtle planes, you can develop an inner ear to hear sounds on the subtle planes. Working with crystals and sound can help you develop your inner ear by which you begin to hear these sounds and become more conscious of the subtle planes which they represent.

In quartz crystal work you use audible, physical sound to change subtle vibrations which, in turn, affect the physical plane. The crystal and sound amplify each other's effect. So, for example, if you want to change an unpeaceful environment to one of harmony, you visualize or concentrate on harmony and create that corresponding sound in the environment. That sound will create the vibrations of harmony in the environment, eventually creating that state. Adding intention behind your sound gives the vibration created enough force to be much stronger than the undesired vibrations. Your new sound vibrations overpower and transmute the other vibrations. The stronger the intention behind the sound, the faster and more effective the change.

You can easily use your crystals with this process because the mechanism for working with sound is largely the same as with crystals. The sound and the crystal actually empower each other, adding to their overall effectiveness. You can work with crystals and sound in the following manner: decide what it is that you wish to accomplish with your crystal. Next, choose the sound or sounds that best seem to correspond with your goal. Then create that sound and send it into the crystal. You can then use your will to send the amplified sound out through the tip of the crystal to do its work, or you can store the sound in the crystal. Sound stored in a crystal causes that crystal to vibrate in a manner which corresponds with that sound. Then you can use the crystal to affect all who come in contact with those particular vibrations. You can actively work with your crystal to later send out stored sound, or you can just leave it in a particular environment to continually radiate out those particular vibrations. You can also wear the crystal or place it with someone else to affect you or them accordingly.

There are many ancient systems for using sound that have been used to heal the body, change the conditions of the physical environment, and to develop consciousness of the subtle planes. This information used to be restricted to shamans, healers and spiritual teachers and their students. Now the information is less restricted as an ever greater number of people are becoming able to utilize this information in a constructive and conscious manner.

Besides the voice, traditional instruments that are used in this manner of sound work are drums, rattles, bells, gongs, cymbals, flutes and dronal instruments such as tambouras and sitars. Each instrument replicates most closely the predominant sound created by the prevalent vibrations of a particular subtle plane. By focusing on one of those instruments, you will start to vibrate in harmony with it and the subtle plane it represents. As you vibrate in harmony with the subtle plane, you can become more conscious of it.

Each particular part of your physical and subtle body is represented by a particular tone. This is true not only of your body, but of every body or form in the universe. A part or whole of any body can either be energized or de-energized when the voice or a particular instrument reproduces that tone.

During work with crystals and sound the voice or instruments are sometimes played alone. Most often they are combined in such a way to effect a particular change in the body and/or consciousness. The combination also can create overtones that cannot be created with a single instrument or voice. These can be especially effective in creating change.

Not only is the particular sound of the instrument important, but also the rhythm in which it is played. Important to the rhythm is not only the sound but also the space in between sounds. Rhythms can represent the breath and/or the heartbeat. By adjusting the rhythm you can change the rhythm of the breath and/or heartbeat thereby affecting the accompanying change in the body, the emotions, the mind or the consciousness. Each plane (astral, mental, causal, etc.) has its own characteristic rhythm that can be reproduced to enable you to breathe or beat in harmony with it. Eventually you may become aware of and perhaps begin to consciously operate in that plane.

When you begin to work with sound you also find that the volume of the sound is as important as the tone and rhythm. Each subtle plane and energy current seems to require its own volume of particular sound. As consciousness flows through various levels, the dynamics of the accompanying sound shift accordingly.

Besides working with specific tones, rhythms, and volume, you can work with a system of sound traditionally called seed sounds. Seed sounds are based on the human voice and are the basic root underlying all language. Each vowel and consonant creates a vibration that is tied to a specific subtle energy current that is located inside the body as well as in the corresponding subtle plane. For example, the vowel sound "AH" corresponds to the subtle energy vortex in the heart center and the etheric element of air. So, the use of the tone "AH" awakens you to the consciousness of that subtle plane which best corresponds with the heart center. Also, you draw into you that air element to affect you accordingly. These seed sounds have been used in the sections of this book which deal with opening the chakras and healing. Seed sounds are combined with each other to produce words or groups of words that produce specific effects on your physical and subtle bodies. They also open your consciousness to the higher planes. These are called "charged words" or mantras. The mantras RAM, SAT NAM, RA MA and WAHE GURU were used to increase the effectiveness in many of the exercises given in this book. Mantras and seed sounds are not only sounded aloud, but are repeated internally to open the subtle hearing. They are often more effective heard with the inner ear.

Thus, when working with sound you can work with tone, rhythm, instrumentation, seed sounds and mantra. The skillful combination of the above aspects of sound into music with the intention of creating certain healing, consciousness raising, or other effects is called "shamanic music." Examples of this are given in two cassette tapes "Helios" and "Wakan Tanka."* These tapes allow you to experience shamanic music and can be used along with all aspects of your crystal work. Helios primarily makes use of gongs and bells. Wakan Tanka uses a combination of drums, rattles, bells, gongs, synthesizers, eastern drone instruments and voice. Both can be used as powerful shamanic tools with your crystal work. They will energize your crystal work and provide a sound stimulus to extend your consciousness into the higher subtle planes. You can use them with crystal healing work as well as visualizations, crystal ball reading and crystal gazing.

* Helios, © 1985, Uma and Ramana Das, U-Music, U-101,
  P.O. Box 31131, San Francisco, California 94131

Wakan-Tanka, © 1985, Ramana Das and Uma, U-Music, U-102,
P.O. Box 31131, San Francisco, California 94131

This has been just a brief explanation of the way sound can be used. There are many systems of using sound, just as there are many systems of working with crystals. One or more of these systems can be memorized and used. However, like crystal working, it is best if your work with it is derived from your own experience. With contemplation, experimentation, and practice, you can arrive at a personal understanding of sound. To truly begin to understand sound you need to be able to still your mind, focus, concentrate, and listen to your own inner voice as you experiment. The more you are conscious of your inner self, the more you can be conscious of subtle sound. As you become more conscious of how to use sound based on your own experience, you can begin to research and try out various ancient systems. Then use what works for you.

How do you choose which sound to use each time you work? What are some exact techniques with which you can begin to work with sound and crystals? The following are some techniques through which the above questions can be answered.

As was explained, each physical sound and its subtle sound counterpart create certain effects on the subtle and physical plane. When you work with sound in your crystal work you must know how to choose a sound based on the effects you know it will produce. You also need to know how loud or soft to make the sound and with what rhythm it needs to be played or sung. To learn how to do this you need only listen to your own inner voice or intuition, just as in crystal work. Developing yourself to be an effective crystal worker will develop your ability to hear and work with sound as well. This exercise wll open up your intuitive abilities to hear sound and know what its subtle and physical effects are. You will then know how and when to use each sound. As with crystal work, the more clear your mind and more focused your concentration, the better your results when working with sound.

SOUND EXERCISE (1)

1. Place a crystal in each of the four directions surrounding you. Visualize a crystal underneath you and over your head as you sit or stand in the center. If you like, hold a crystal in each hand. Then place crystals over your throat center and/or third eye point. (This can be done with a necklace or headband.) These will help you to have the energy to focus clearly and concentrate on each sound and its effect. This will also help to open your capabilities to be aware of the subtle sound which accompanies the physical sound.

2. Center and ground yourself. Calm and clear your mind. If you like, use techniques given in this book to do this.

3. Close your eyes. Do not focus on anything in particular. However, during the course of this exercise the focus of your eyes may shift to your third eye point. If that happens, let your focus remain there.

4. Find the note that is easiest and most comfortable for you to sing and makes you feel the best. (This is your key note.) What note is this? Concentrate deeply on this note as you hold its tone as long as you are comfortably able. As you sing this note, notice its effect on your various chakras. What energy center does this sound seem to affect most deeply?

What is its effect on that center? How does it affect your body? Are there any particular parts of your body it affects more than others? How does it affect these parts?

5. What kind of energy or life force does this sound seem to summon?

6. Focus on your emotions as you continue to sing this note. What emotional response does it seem to create? Does it continually seem to create this effect?

7. Next, focus on your mental state as you sing this tone. How does it seem to affect your overall mental

state? Are there any particular thoughts or images that it seems to engender?

8. Does this sound seem to give you extra energy? What type of energy does it seem to give you? Does this tone seem to calm you or fire you up?

9. How far does this sound seem to draw you into a harmonious or rapturous state? (If you like you can rate it on a scale of one to ten, ten equaling highest rapture.)

10. How does this sound heal?

11. Next, experiment with the dynamics of the sound. What happens if you sing this sound softly? What changes do you note as you sing it progressively louder?

12. Experiment with the rhythm of the tone. Pulse it in different rates of time. Note any differences in how you feel as you alter the rhythm. What seems to be a rhythm that especially fits this sound?

13. Next, experiment with different seed sounds. Sing each vowel sound and note any differences in how you feel. What vowel sound seems to best fit this tone?

14. Next, continue this exercise, changing from tone to tone. (The best method is to go up the scale in whole, half or quarter tones from your keynote. Then go down the scale in whole, half or quarter tones.) As you focus on each new tone ask the questions that you asked of the first one. Cover as many tones as you like. Does there seem to be any general tendency as you sing progressively higher or lower tones?

After working with tone and seed sound, experiment with various instruments. Again, working with yourself as the laboratory, ask yourself the following questions as well as any other which might occur to you.

15. Sit with one or several varieties of drum. Begin to beat the drum in one particular rhythm that first seems to appeal to you. Ask yourself the above questions that you did before, with the exception of #13. Instead of that question, ask yourself how the drum(s) is best played with each seed sound. How do you feel as you add the drum to this sound?

16. After the drums, experiment with rattles, gongs, bells, cymbals, flutes, etc. If you do not have these available to you, listen to recordings and investigate as best you can.

17. After you are through with the exercise, clear your crystals and store them. Clear yourself and the room you are in. Ground yourself again.

Listen and investigate only one or two tones or instruments at each sitting, or as much as your concentration will allow. If you tire yourself you will not be able to hear as well and cannot investigate as deeply or as accurately. This process takes time and cannot be done effectively in only one or two days. Be patient. The point is not how rapidly you investigate, but how deeply and effectively. As you do this exercise you will develop the ability to automatically be conscious of the effects on your physical and subtle bodies of any sound you hear, at the time that you are hearing it. This will give you the ability to play shamanic music and use sound to heal, change consciousness and otherwise affect the physical and subtle bodies and environment of yourself and others.

Doing the prior sound exercise will create in you a mastery in working with sound and crystals as long as you continue to use your "intuitive ear" rather than an intellectual/analytical ear. However, you can begin to work with crystals and sound without doing the prior exercise as long as you still rely on your "intuitive ear." The mastery created by the above exercise will allow you to use sound with more precision and more consistent, dependable effectiveness. It is possible, however, to create results just by relying on your intuition in each particular situation without the background of experience sound Exercise #1 gives you. However, your results may be less specific and less dependable. The following is an explanation of how to choose which sound to use in a particular situation along with some specific crystal techniques to use.

USING SOUND TO AUGMENT YOUR CRYSTAL WORK

1. Choose what it is you would like to accomplish with the use of your crystal(s).
2. Begin to use a crystal method that best fits what you would like to accomplish.
3. As you use the crystal(s) in the particular situation, intuitively choose a sound which best reflects what it is you wish to accomplish. For example, if you are doing a healing technique with your crystals, focus on the healed condition as if it had already happened. What sound or series of sounds seems to best fit that healed state?
4. Next, sing that one or play it on an instrument.
5. Send that sound in and through the crystals and out its top. This augments the effect of the crystals.
6. If you like, you can program a crystal with sound by creating a sound or sounds and sending it in, not through, the crystal. If you like, lock the sound into the crystal using the technique given to program a crystal. The crystal will then radiate vibrations that correspond to the sound until you clear it. Have the intended recipient of the sound vibrations wear the crystal, keep it with them, or keep it in their environment.
7. After you have finished working with the crystals, clear them. This will clear any residual sound vibration out of them as well as anything else you perhaps pulled into the stone.

How do you choose which sound to use? If you have done Exercise #1, refer to your backlog of experience and refer to your intuitive mind to choose what best seems to match the situation based on your past experience as well as your immediate sense. If you have not done Exercise #1, intuitively choose a sound which best fits the immediate situation. Either way, ultimately rely on your intuitive sense. Once you have chosen a sound, let go of any interfering judgmental thoughts or feelings and use it. Continue to use the sound, varying it as necessary, until you feel that it is time to stop. Later, check your results in the physical univertse.

GEOMETRIC FORMATIONS USING QUARTZ CRYSTALS

Every geometric formation creates a like arrangement of the vibrational energy patterns in the subtle and inter-connected physical plane. Certain attributes, activities and powers are activated by these energy patterns created by the geometric form. These are available to be worked with and can add to the effectiveness of your crystal work.

What are some of the more traditional geometric shapes which are used?

The most familiar is the circle, often associated with protection and centering. It represents or gives a sense of infinity or the endless circle of changes, etc. There is the white circle, which in some traditions represents the etheric element.

The square is another familiar form. The square can be used to call upon the powers of the four directions. East traditionally represents spiritual enlightenment. The color associated with this is yellow. South represents death, endings or change. The associated color is red. West represents the great mystery, the void, or the unknown. The associated color is black. North represents healing. The corresponding color is white. If you work with six directions, blue represents the sky overhead and its energy. Green represents earth and its energy. The yellow square in some traditions represents the element Earth.

The triangle is also common. The red triangle can be seen as representing the fire element. Another common representation is the trinity. The triangle can be used for sending messages out of the tip.

The crescent moon with its lunar feminine aspects, or the white crescent representing the water element is a familiar shape. The five-pointed star represents the five elements or the perfected man. The six-pointed star reflects the union of the spiritual and the material, or of man and God. The lists and systems regarding sacred geometrical formations are many. To include the use of geometry in your crystal work you need to know them not just intellectually, but also to experience their energies. The best way to experience the energy associated with a geometric space is to set it up around you and intuitively sense what it feels like.

How do you set these spaces up? There are many methods. This method is done with the use of quartz crystals and focused visualizations. When you use visualization, depending on the clarity and strength of focus, a thought form of the visualized shape is created on the mental and astral planes. As explained before, the thought form arranges the mental and astral energy patterns which then similarly affect the physical. (Whether or not an actual physical form is created.) The quartz crystals can amplify the energy patterns on the astral and mental planes, causing them to vibrate more intensely. This, in turn, makes the resulting physical effects of the geometric creation stronger and more effective.

The following techniques show how to work with quartz crystals in order to set up these geometric shapes and to experience their power. The square will be used in this example. You can use the same technique to experience other shapes. The materials you will need for this are four good-sized quartz crystals, a hand-held crystal or crystal wand, and possibly one or two other crystals that are of similar size and/or energy. To start, you probably will want to use clear crystals rather than smoky or amethyst so you won't be limited to the use of the violet or brown color in your work. The wand crystal or single crystal that you will use like a wand should only have one point and not be double-terminated.

GEOMETRIC VISUALIZATON TECHNIQUE

1. Before beginning the formation, sit quietly in a space where you will not be disturbed. Lay your crystals in front of you for future use. Sit in a relaxed manner with your spine straight and begin to breathe long deep

breaths in and out of your nose. Gently fill your lungs completely and them empty them. Peacefully allow your thoughts to come and go, not attaching yourself to any of them. Pay attention to the breath. If you find that your thoughts have wandered, just drop what you were thinking and bring your attention back to the breath. You will feel yourself relaxing and feel your mind becoming calm and clear.

2. Next, drop your attention from the breath and retain your calm, clear focus. Begin setting out the square around you in the following manner: Visualize around you a square of any size with you in the center. Hold the visualization while you take the first crystal and place it to mark the right corner of the imagined square.

3. Take another crystal and place it to mark the next corner in back of you.

4. Continue to mark all four corners of the square with the quartz crystals. Point them in the direction that you feel drawn to. You can go clockwise or counterclockwise as you desire.

5. Next, as you stand in the center of the square, use your crystal wand or single-terminated crystal to visualize "drawing" a connecting line of energy between each corner. This connecting line may be drawn with the tip of the crystal on the ground, or may be directed from where you stand. Visualize this line as gold, silver, violet, or any color of which you would like to use the energy.

6. When you are satisfied that the square is complete, sit or stand in the center. Continue to maintain the clear, focused state of mind with which you formed the square.

7. Next, allow yourself to open up and focus on the energy created. Hold a quartz crystal in each hand to help you attune yourself to the formation. With your mind focused and all intellect aside, notice what it feels like inside the square. What does your body feel like? Are there any images or thoughts that seem to be connected with the square? If you haven't programmed any specific

color, what color does it seem to be? Are there any sounds that seem to be connected with the form? Does there seem to be any change in temperature? Notice everything -- become familiar with that particular form.

8. Next, focus on the energies associated with the four directions. First, face the West, the direction of difficult lessons to be learned, things to be overcome; the unknown. Picture it black. Use your will to draw that energy into you, if you need to use it or simply experience it. Going clockwise, face the North, the direction of healing and renewal, the associated color white. Let the energy of the North wash over you. Next face the East, the direction of enlightenment, the color yellow. Call for enlightenment, and listen to anything that you seem to hear in your intuitive voice. Finally, face the South, the color red, the direction of beginnings and endings, birth and death. All of the directions offer their particular energy, their particular wisdom and help -- as does the entire square itself.

9. Finally, stand in the center of the square and listen to your inner voice for any direction, empowerment or further suggestion as to other attributes that shape may have. These are only some suggestions. Follow your feelings. Be still, be attentive and be conscious. The teachings will come. Use them wisely.

10. After being in the center of the square you will eventually want to leave. Allow yourself to withdraw from the square. Then begin to dismantle it. First, taking your crystal wand or hand crystal, "undraw" the connecting line between the crystal corners of the square.

11. When that has been done, one by one, in the opposite direction in which they were laid out, remove the corner crystals. Place all your crystals in front of you.

12. Then, become aware of the room environment around you and take a couple of deep breaths and expel them sharply. You may want to shake your body. Then

sit or stand for a couple of minutes feeling the earth or the floor beneath you. Feel like you have roots deep within the earth from the bottoms of your feet.

13. Finally, clear the crystals, yourself and the environment or room about you with any method with which you are familiar. You may want to record your experiences for later integration and use.

This method can be used to familiarize yourself with as many geometric shapes as you can imagine. In Step #8, face as many directions as the shape has and experience their energy. As was mentioned earlier, there are many systems of meaning for the various geometric forms. It is useful to learn what they are but for the most effective work with them, you must experience the meaning yourself. Trust the inner wisdom which flows through you.

When working within a geometric shape by the above method, you are focusing and working with only one dimension of the form. In actuality any geometric form is infinitely dimensional. Each dimension extends infinitely in a set pattern triggered by the original geometric pattern that you created. In other words, if you stand in the center of a square shape that you have created on the ground you actually have a square shape standing up in front of you, in back of you and at each side as well. You are standing in a box. This box does not not just surround you with six sides like a cube suspended in space. The lines creating the square extend infinitely to create square upon square; you have actually created an entire universe of squares in which you stand at the center.

Space, beingness or "is-ness" contains all possible or potential geometric patterns as it contains all dimension and pattern. Each geometric pattern extends through space to create a gridwork of endless replications of the original form. These gridworks are all contained within one another. Each grid creates a particular path through space. This geometric gridwork vibrates in a particular pattern which can be felt as you focus on the single geometric shape. You can train yourself to be sensitive to the geometric shape in the exercise above. Then you can extend your consciousness and use your will to "travel" through space using the lines of the grid created by the original geometric shape. The exact center of the multi-dimensional geometric grid system is where manifestation and demanifestation take place.

✑ You are always the center

COLOR AND COLOR STONES

In every aspect of your crystal work you will find color a useful tool to work with. The addition of color to your work can be effected with the use of color stones and other colored materials as well as with visualization techniques.

Every color has a certain appearance which can evoke certain emotional and psychological reactions. Also, every color has its own particular type of vibration which can be amplified, stored and transmitted with the use of the crystal. Any color which is placed in physical or visualized association with the crystal automatically causes the crystal to vibrate harmoniously with the color. That crystal vibration, in turn, can affect your physical, etheric, emotional, and mental body as well as your environment. This automatic process can be amplified even further when you use your focused will to intend for it to happen.

There are many systems of working wth color. Also, there are many ways to work with colored stones and crystals. Many methods which can utilize and incorporate color have been used throughout this book, particularly in the chapters on healing the extended body, and crystal jewelry. As with the rest of your crystal work, you can memorize such methods and put them to use. However, it is most effective to learn how to sensitize yourself to color and then use it on the basis of your own experience. Each situation in which you do crystal work is unique, requiring you to use your own inner wisdom and sensitivity to act in the most appropriate manner. You not only use your clear crystals differently each time, but color and color stones as well. If you have to use a particular system rather than your own sensitivity, you do not have the ability to respond uniquely in each occasion. You then work in terms of generalities rather than specifics. For example, if you know that a green or pink stone is good for the heart center you will not be able to intuit a time that the heart area seems to need cooling and calls for a pale blue. Or that the lungs in that general area may best be helped by a yellow color. Instead you will automatically use pink or green. This will not be harmful and the center itself may be stimulated, but it won't help as much as if you were able to be sensitive to the exact situation and the exact color to use.

When you work with color, be sensitive to not only the color, but to the shade of color. Each has its different quality. When using color stones be sensitive to the degree of opaqueness, brilliance and clarity involved. Generally the more clear the stone the more etheric quality it has. An opaque stone has a more dense, grounding quality. Each color has a sense of a different degree of coolness or heat. Some colors feel more harsh or abrasive in quality while some are more soothing and subtle. When working with color, stones and objects, notice that each has a subtle difference even if they are the same type or color. These differences cannot be seen, only sensed.

Colors are often used in combination with each other in crystal work. Again, you must be able to sense the result of each combination rather than relying on your intellect as usual. For example, if you combine a yellow color with a red color, your logical mind would suppose that the resultant effect would be that of orange. Possibly. However, the effect of red and yellow can be quite different than that of orange. For example, an orange stone can work on your second chakra, or your sexual center. A garnet used with a yellow citrine can stimulate your first chakra, perhaps giving you grounding, while the yellow citrine stimulates your navel area. The total effect of that would be to enable you to manifest on the physical plane while remaining grounded. This is entirely different than stimulating the sexual center.

DEVELOPING SENSITIVITY TO COLOR

1. First, work with one color at a time.

2. Sit or stand in a quiet room where you will not be disturbed. Either fill the room with a particular colored light or gaze upon an object of particular color. If you like, you can gaze at a large piece of colored paper.

3. Center yourself and clear your mind.

4. As you gaze upon that color or look at it around you in the room, notice how you feel. Notice the obvious effects as well as the very subtle ones.

5. Ask yourself these questions:

    A. What emotional state does this color seem to create in you?

    B. What thoughts do you seem to have in response to it? Or what state of mind does it seem to cause in you?

    C. Do you feel more excited and active, or more calm? To what degree?

    D. Describe to yourself all of the qualities of this color which you sense or feel.

    E. What parts of your body does this color seem to affect?

    F. Based on these prior observations, how would you generally tend to use this color?

6. Finish looking at the color or turn the colored light off. Clear and ground yourself.

7. Continue this process several times with each color.

8. Do this with as many colors as you want to or can think of. Then work with combinations of colors. Sense their relations with each other. Answer the questions in #5 above with each color combination.

9. Then, start wearing clothes of one color for a while. At the same time do as much as you can to have your environment reflect that color. Each day notice how

the color is affecting you. Notice each part of your body that seems to be affected.

10. You will be able to experience the full effects of this color if you do this for 30 days. Try this with every color.

11. You can also sensitize yourself to color if you wear white clothes for an extended period of time (at least 30 days). Then try wearing different colors of clothes, one at a time. After wearing the white, you will be particularly sensitive to other colors and their combinations.

DEVELOPING SENSITIVITY TO COLOR STONES

The process of becoming sensitive to color stones is much the same as the process of becoming sensitive to color itself. When you become aware of the qualities of color and their possible uses, you have explored only one aspect of color stones. They also have properties of brilliance, clarity and cut that must be considered. To become sensitive to a color stone, do the following:

1. Clear your mind, center and ground yourself.

2. Concentrate on the stone as you hold it in your hand(s) and consider the following questions:

3. How does the color affect you? Ask yourself all the questions in #5 of the first process.

4. Does the stone feel cool or warm, irrespective of its color alone?

5. How dense does it feel and how does that affect you?

6. What other vibrations seem to be stored in the stone? Focus on your third eye as you hold the stone to it. What images come to mind? What is the history involved with this stone and how does it affect you?

> 7. What is the cut of the stone? How does it seem to affect the other qualities of the stone?
> 8. What other qualities do you intuit about this stone?
> 9. Put the stone down, ground and clear yourself.

Keep records of your observations. You will be able to refer to them for future use. You will notice generalities of the stone that seem to hold true time after time. Other observations will apply only to a particular stone or a unique situation.

These prior exercises will yield not only information about colors and color stones, but will develop your ability to quickly, accurately and easily intuit qualities and uses of color and color stones in every situation. This will allow you to use them with complete effectiveness in your crystal work.

Once you have developed the ability to quickly and accurately sense the properties of color and colored stones, how do you use this skill in your crystal work? Basically, you use color along with the quartz crystals to re-create the original balance and harmony in any body or situation with which you are working. The technique is the same as when you are using sound or projecting emotional or mental states and visualization. You first see that which needs to be changed. Focus and sense the accompanying vibration that accompanies that state. Sense the color which best corresponds to the vibrational state. Next, use your will and concentration to change the vibrational state to a better one. As you do that, sense the color that best corresponds to the desired result. Project that color as you use your will to change the vibration. In your mind's eye, see the old color change to the new color. This will further increase the accuracy of and hasten the change process.

If this is unclear to you, the following example of curing a headache in another person may be helpful.

CURING HEADACHES

1. Sense the accompanying vibration to the headache. How does it feel?
2. Visualize or intuit the color associated with the feeling of the headache.

3. Next, sense or visualize the feeling or vibration of a head free of tension and pain.

4. Sense what color corresponds to that which you are feeling in #3.

5. Will or strongly intend that the feeling or vibration of the headache change to the new feeling of a head free of pain and send it out into the person's head.

6. As you do that in #5, visualize the new color which corresponds to the feeling of a head free of pain and send it out into the person's head. See it replace the old color.

7. To send the color to the person's head, you can place a similarly colored stone on their forehead. Or you can transmit the color from a color-filled crystal that you hold in your hand. You can also put colored light, cloth or any other material on their head. Instead of using a color stone, crystal, light or other object, you can rely on visualization. (Remember, the effectiveness of your visualization is only as strong as your power of concentration and strength of will.)

8. If you want to continue to affect this area without being present yourself, have the other person wear the color stone(s) directly on their forehead or in the vicinity of their head. Or program a clear crystal to act as a certain color and place it on or near their head.

9. Be sure to clear the stones, yourself and the other person when you are finished.

As you can see, working with color stones and color uses many of the same principles and techniques as working with clear crystal. You need to be sensitive, spontaneous, strong-willed, concentrated, and reliant on your own inner wisdom.

Though it has been explained to you how to select and use colored stones, a chart follows to help you get started. You can do all your work with the colored stones listed above. However, if you want to work with or are curious about another stone, just include it in the most similar category of the above chart. This will give you an idea of some of the qualities of the stone.

General qualities and uses for each type of stone are listed. The part of the subtle and physical body that corresponds with each type of stone are also included in the chart. Each type of stone is used to do work in the corresponding subtle and physical body areas. However, the stones do not have to be used only in that body area. They can be worn or used anywhere in the body to influence it with the stones' qualities.

Use the stones listed above in the methods described in the text. As you continue to work with each stone you will begin to notice more particular effects associated with each individual stone beyond what is shown in the chart. As explained in the book, each circumstance in which you work with your stones is different so this chart can only serve as a general guideline. To learn more, work with your stones!

COLOR STONE CHART

Color Group	Subtle Energy Location	Approximate Location in the Physical Body	Types of Stones	Qualities of the Stones	Some Uses of the Stones (in the corresponding physical or subtle center or when put in contact with the body as a whole)
Black	Center of the earth	Bottoms of feet	Obsidian, Onyx, Black jade, Agate, Black coral, Iron pyrite	Cold, solidity, stillness, sleep/rest, no movement, coolness	To open energy channels from bottom of feet. Grounding and focusing. Calming the mind and emotions. Quick, firm focus on the physical plane. Slowing/stopping subtle or physical movement.
Grey	Middle of the earth	Lower leg	Smoky quartz, Obsidian, Grey pearl, Agate	Same as above but with less firmness and slightly more movement	Same as above, but not as extreme. Opens energy channels down legs to earth.
Brown	Surface of the earth	Upper leg	Smoky quartz, Obsidian, Brown topaz, Brown tiger eye, Jasper, Agate	Peacefulness, nurturance, safety, warmth calm, placidity	To lightly ground and focus the mind. Calms the emotions. Soothing to the physical and/or subtle bodies. Opens energy channels down.
Red	1st chakra, root center, base of spine	Base of the spine and physical heart, blood, circulation	Garnet, Ruby, Coral, Jasper	Warmth, fire heat, high-action	To open the 1st chakra. To break through mental conditioning. To create rapid change/empower action to energize. To create heat. To work with heart, blood, circulation.
Pink	The heart center	Inside the center of the chest, the heart, any area that needs gentle change	Rose quartz, Pink tourmaline, Rhodochrosite, Pink coral, Pink jade, Pink garnet	Gentleness, love, peace, warmth	Lifts depression, creates peaceful feelings. To gently energize. To create warmth. To relieve stress and disharmony. To open the heart chakra. To open the energy channel between the 1st chakra and heart center.

156 / *The Complete Crystal Guidebook*

COLOR STONE CHART

Color Group	Subtle Energy Location	Approximate Location in the Physical Body	Types of Stones	Qualities of the Stones	Some Uses of the Stones (in the corresponding physical or subtle center or when put in contact with the body as a whole)
Orange	2nd chakra	Near the sexual organs, pelvic area	Carnelian, Agate, Madiera citrine	Aroused sexual energy, warmth, activity, fire	To lightly energize. To increase creative energy. To increase sexual energy. To strengthen and stimulate 2nd chakra.
Gold or yellow	Gold - the energy channel between your 3rd chakra and through your crown center. Yellow - 3rd chakra	Near the navel area to the top of the head	Yellow citrine, Light smoky quartz, yellow diamond	Warmth, activity, sunlight, creative energy	Increasing male energy. To manifest creative energy on the physical plane. Lifting depression. To lightly energize physical body and subtle nervous system. To work on lungs, breath and life force. To strengthen and stimulate 3rd chakra. Improving force of will.
Green	Heart center or 4th chakra	The center of the chest, any area that needs cooling, lungs	Emerald, Green malachite, Peridot, Tourmaline, Green dioptase, Jade, Green calcite	Generosity, prosperity, nurturing, cooling, slightly calming, water	Relaxing and expanding the heart center. Cooling fevers. Calming heated emotions. Increasing prosperity consciousness. Relaxing cramped muscles and tightness in the body. Strengthen the physical heart. Increases the ability to love.
Turquoise sky, or light blue	5th chakra throat center	Center of the throat, part of jaw and ears, back of neck, base of skull	Turquoise, Gem chrsyopase, Aquamarine, Blue topaz, Blue malachite, Blue quartz, Celestite, Blue tourmaline	Sky, lightness, vibrant repose, joy, astral	Astral or out-of-body travel. Relaxing stress, particularly in the jaw, neck and top of shoulders. Increasing verbalization and singing agilities. Increasing effectiveness of communication. Opens subtle energy channel from the heart to the third eye. Joins love with wisdom, creating compassion. Getting rid of headaches, calming fever.
Royal blue to indigo	6th chakra third eye center	The center of the forehead, part of ears, eyes, and upper head	Sapphire, Lapis lazuli, Sodalite, Dark blue Tourmaline	Regality, deep space, endless repose, calm strength, spiritual	Calming the mind. Becoming more intuitive. Developing wisdom. Increasing mental abilities. Focusing and maintaining concentration. Improving memory. Improving eyesight. Stilling the body. Becoming meditative. Increase sensitivity to subtle energy.

COLOR STONE CHART

Color Group	Subtle Energy Location	Approximate Location in the Physical Body	Types of Stones	Qualities of the Stones	Some Uses of the Stones (in the corresponding physical or subtle center or when put in contact with the body as a whole)
Deep purple to light violet	7th chakra crown center	Center of top of head, slightly above head	Amethyst Violet tourmaline	Calmness, gentleness, expansive, regality, harmonizing, spiritual	Increasing spirituality, stimulating enlightenment. Physical, mental, emotional subtle healing. Lifting depression. Relaxing stress. Increasing feminine energy. Dropping unwanted desires. Developing calmness in body, mind and emotions.
White	Higher energy centers than the crown center	Above the top of the head	Clear quartz White diamond Pearl Calcite	Clarity, brilliance, light, ice, etheric, endless potential, ever changing	Clear quartz, diamonds and sometimes calcite can be programmed to take on the characteristics and abilities of any color stone. They contain all color so have all possible uses available to them. They can energize, harmonize, heal and expand the capabilities of all bodies. Calcite and pearl can be used to strengthen and stimulate bones, hair and nails.

Note: Opal can be included in the white category as it contains all colors. The difference is that opal has the quality of fire and so is ever changing. It stimulates activity and change. If the opal has more of any one color, the qualities of that color are dominant along with the changeability of the fire. Only stable people should wear them.

CRYSTAL TOOLS

In nearly every great culture of the past, quartz crystal tools have been fashioned and utilized -- in the form of wands or sceptres, meditation objects, healing objects, power pieces and ceremonial jewelry. The information about and use of these tools was restricted to priests, mystics, shamans, healers and their initiates. Today, various methodologies have been "rediscovered" and have become available to the interested person.

Fashioning today's crystal tools is a creative challenge, as correct placement, understanding of the crystal, and application of artistic design and beauty is involved. When selecting your tool to use or wear, there are certain features to look for beyond its basic artistry. How is the crystal attached? With what material(s) is it joined? Is the metal or other material appropriate for you? Does the placement of the crystal on your body strengthen you? Does a piece of crystal jewelry lie on your body near the area you want to have "open," healed, or energized? If there are other color stones included in the design, are they right for you? Does the crystal tool feel balanced?

Any object attached to the crystal adds aspects of itself to the crystal tool. For example, bone will add more of the earth element. Also, the nature of the creature from which the bone came will influence the overall tool. Gemstones have specific qualities due to difference in color and vibration, so that combining gemstones with crystal will affect the overall piece. Feathers, often attached to ceremonial or shamanistic tools, connect the nature and essence of the particular bird to the piece.

The crystal should not be attached in such a way as to depend entirely on glue, epoxy or cement. If it does, the crystal will probably fall out at some time. The crystal should not be completely drilled through horizontally (from side to side) because that interferes with the energy flow. The top of the crystal should not be completely capped, to allow for the crystal's tendency to slightly expand and contract as it becomes energized.

Observe the design used in any tool or jewelry piece. Certain symbols can be used that will affect the crystal with their vibration. For example, circles, squares, triangles and other geometric shapes have particular energies and meanings. Symbols such as stars may vary in their energy depending on the number of points to the star and the direction of the point in relation to the crystal. Yantras (meditative designs) and other archetypal symbology can be combined with the crystal to create powerful meditation and healing tools. Seed sounds or words used in mantras can also be employed in a similar manner by applying the written sound or word to the crystal piece. Of course specific objects, animals or mythological beings can also be integrated successfully to influence the crystal with their energy. For example, the serpent or snake often symbolizes the rise of the creative empowering, expanding kundalini energy. This design is particularly effective in a wand or tool intended to raise energy to direct out through the crystal portion of the tool.

The above are a few of the many aspects of making crystals into powerful tools. If you are feeling drawn to the tool, object or jewelry, ask about the materials and symbols employed. Then listen to your inner voice. You will know if it is right for you. Through guided intuition, you will also learn how to best use it. Conscious use of crystal tools is an interactive process, a two-way channel between you and your tool. You direct each other. In this way, the potential of the tool can be realized, making available heightened awareness, energy, healing, magic, vision and empowerment.

The following pages describe in more detail the types of jewelry, power and altar objects that can be made with quartz crystals. Discussed are their different uses, information on metals, setting considerations, color stones, storage and balance. Finally, information is included about the different types of symbology that is often utilized in the design of quartz crystal tools.

WHERE QUARTZ CRYSTAL JEWELRY
IS WORN

Quartz crystal jewelry is generally worn over the energy meridians and near the chakra points of the body. Crystals worn at these areas will tend to open and increase the flow of energy in that particular area that is in direct contact and in the immediate vicinity of the stone. Of course, since even a small crystal will radiate an energy field of at least three feet, all crystal jewelry will tend to energize and/or heal the body as a whole. The strongest influence, though, is that area closest to the crystal itself.

Crystal necklaces are worn at the throat center or may hang over the heart center or somewhere in between. Those worn at the throat will open and make available the associated attributes. Those worn over the heart will open the heart chakra and do healing work on the circulatory system and physical heart itself. A pendant worn between the throat and heart will work on both of these areas as well as provide healing for the lungs and respiratory system. When the tip is pointing down, it tends to provide a grounding influence as well as energizing the body. This is often more balancing. When the tip points up it tends to direct more energy to the higher chakras as it energizes the body. If you tend to be absent-minded or "spaced out," this could be too ungrounding for you. A double-terminated crystal focuses equal amounts of energy to the higher and lower centers. A double-terminated crystal pointing side to side tends to focus the main energy to the center of your body on the level that the crystal rests. A crystal worn over your heart center will help deflect negativity and provide balance for your body energies. This balance will help you to relax.

Crystal bracelets work on the meridian points in the hands and also work on channeling energy in and out of the arms and hands. Crystal bracelets are excellent for healers, bodyworkers and others who work with their hands. It is particularly effective and balancing for the body to wear a pair of crystal bracelets that are equally matched or balanced in terms of the size and quality of the crystals and metals around the stone. When a crystal bracelet is worn on the left wrist, point the crystal toward the shoulder.

The left side is the receptive side that generally pulls energy into the body. The crystal pointing toward you will aid this process. On your right wrist, wear the crystal pointing out toward the fingers to aid the process of the out-channeling of the right side. This use of crystals will help balance your sun/moon or male/female energy as well as increase the flow through your hands.

Crystal rings `will help open the energy channels at the end of your fingers. Each particular finger will have a different effect as each is associated with a particular type of energy. There are several systems showing the energy associated with each finger. In one system, the fifth finger is associated with psychic energy, the fourth finger is associated with sun energy or vitality, and the third finger is associated with will power or Saturn energy. The second finger is associated with wisdom. The thumb is associated with your own personal ego. (Generally, you do not wear a ring on that finger.) Rings worn on more than one finger combine these energies. As in the case with the use of bracelets, the right side channels energy out from yourself and the left channels energy into yourself.

Crystal earrings are worn on or very close to the actual acupuncture points that work directly on the third eye. Therefore, they are very stimulating for this center and increase your intuitive and psychic powers. They will also work with the entire head area to provide healing for the sinuses, the ears and the eyes. Again, crystal earrings worn on your left side will channel energy into you and focus healing on the left side. Crystal earrings worn on the right side focus healing on the right and channel energy out from you. You can work with them to balance the two sides of the brain. An earring will increase the energy of the hemisphere that it lies nearest.

Crystal headbands place the stone directly over the third eye point, so will even more directly work to open the third eye and its various powers.

Crowns and headpieces help open the crown center, the energy channel that opens you to the highest consciousness that then floods through your body.

Crystal belts and waistbands help to open and stimulate your navel center. This helps discharge repressed anger and strengthens the nervous system. If your higher centers are open, it will help to balance your body energies.

Crystal anklets can be worn to help you remain grounded. They will also open the energy meridians on the bottom of your foot as well as the channels running down your legs. Rings worn on the toes will open the energy channels out of the ends of your toes. As with fingers, each toe is connected with a particular form of energy.

Wands amplify the energy of the stones and create a laser-like beam of energy that can be directed for certain effects. A wand can be used in conjunction with your intention to cut through, disperse, form, remove and otherwise direct subtle energy in an extremely powerful and precise way. They are used in healing, psychic surgery, ritual and any other situation that requires a precise powerful flow of energy. They are also admired as beautiful sculptural objects. The crystals in a wand should always be pointing in only one direction so that the energy flow is amplified in that direction. If there is a stone in the bottom that points away from the tip, it will detract from the strength of the energy flow. It may even prevent it. A stone pointed side to side on the bottom of the wand will not interfere with the uni-directional flow of energy.

Dorjes are similar to wands but have crystals pointing in opposite directions at each end. The crystals on each end must be the same strength and size. They also are used for directing energy, often in conjunction with hand positions or with sound. An energy source is generated apart from the dorje. Then the dorje spins, reverses and otherwise directs the flow. The dorje can be used to represent the spine or the kundalini flow of energy up the central cord. It can be harmonized with this energy and then manipulated to adjust, unblock, amplify, or otherwise change this flow.

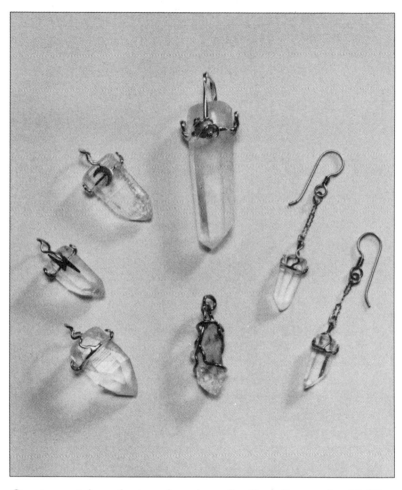

Quartz crystal pendants and earrings. Sterling silver and gold-fill metal.

Four styles of natural quartz crystal ball pendants. Sterling silver.

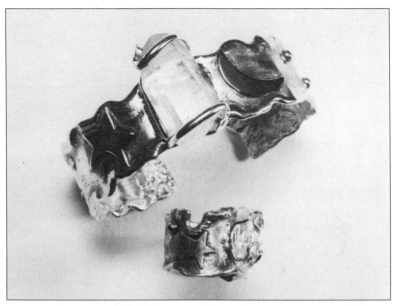

"Cosmic" design for bracelet and ring. Mixed metals.

Custom bracelets in sterling silver.

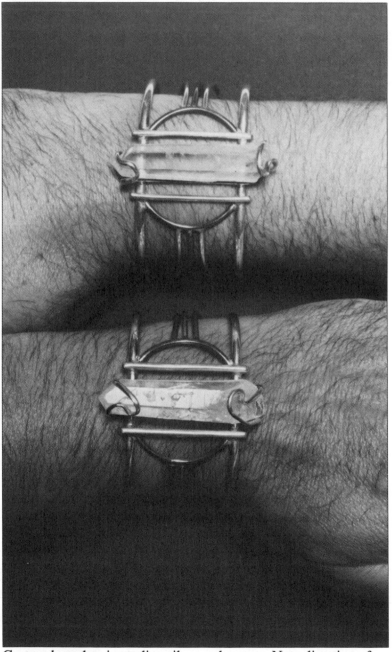

Custom bracelets in sterling silver and copper. Note direction of crystal's flow.

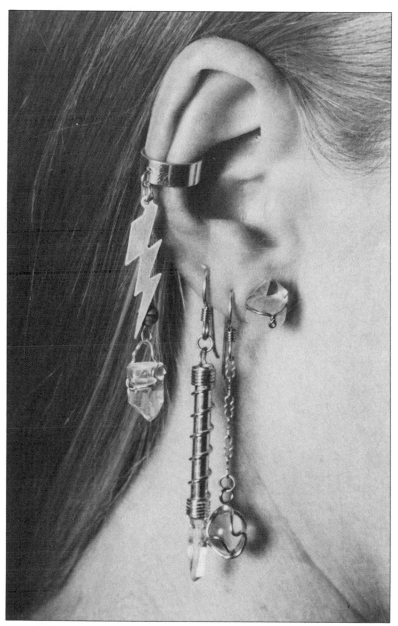

Quartz crystal earrings, ear studs and ear cuff.

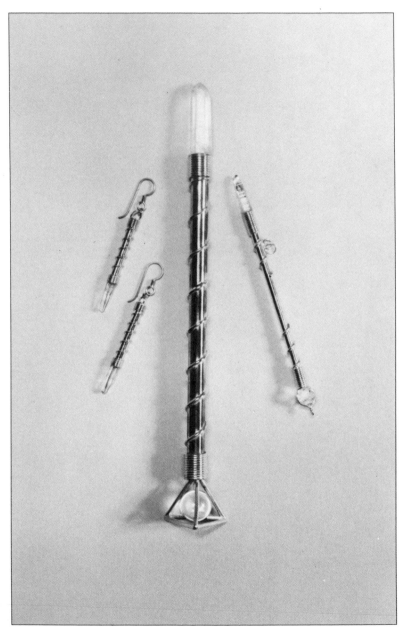

Wands and wand earrings.

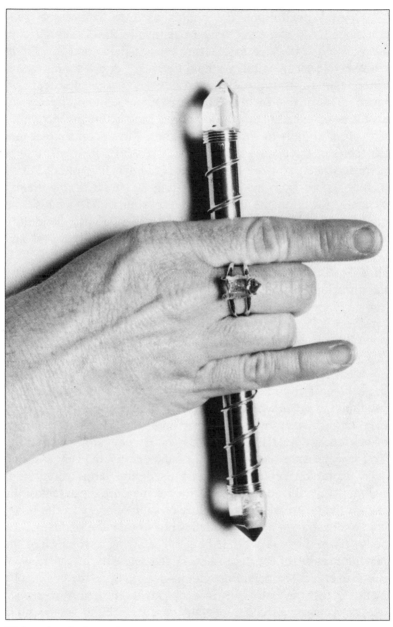

Using the dorje with quartz crystal ring.

Crystal pendulums are   tools used to reflect the sub-conscious inner awareness which continually flows through you. They also can be used to measure energy fields and the subtle currents of energy within your subtle body. A pendulum uses a quartz crystal pointing downward which is suspended from an approximately five-inch length of chain or cord. The crystal should be balanced so that the top points exactly downwards rather than tilting slightly to one side or another. Most people prefer that the facets of the crystal are as even as possbile, feeling that it is more accurate. To work a pendulum, hold the chain or cord three to four inches from the crystal, letting the stone dangle freely. Mentally focus on the crystal and use your mind to cause the crystal to rotate in different directions. See how the pendulum reflects your mind. Now, think of the word or answer "yes" and see which way the pendulum rotates. Then focus on "no." It should rotate in the opposite direction. Once you have determined the direction of rotation (which is usually clockwise for "yes"), you can begin to use it to answer any inquiry which requires a "yes" or "no" answer. Ask your question and then keep your mind "blank" until you receive your reply which is denoted by the directon of the pendulum's rotation.

You can use the pendulum to measure the strength or projection of a field of energy. Mentally direct the crystal pendulum to rotate when it detects a field of energy. The stronger the rotation, the stronger the energy field. You can find the center of any energy field by finding the maximum point of rotation. You can measure the projection of the energy field by walking away from the center until your pendulum stops rotating or swinging. Do this in many directions to measure the extent of the energy field. To measure the flow of subtle energy in the body, hold the pendulum about six inches higher than the surface of the body. Then move it over the body. The swing or rotation of the pendulum will reflect the direction and intensity of the flow of subtle energy. The pendulum can be used to open the chakras by mentally causing it to spin in a clockwise direction over each chakra. To close a chakra, mentally cause the pendulum to spin in a counterclockwise direction. The best way to become familiar with the many uses of a pendulum is to experiment with it. The more you use it, the more your accuracy will improve. As with other crystal jewelry, make sure that the crystal, the metal or other materials used in its construction, as well as any symbology or added color stones are right for you to have.

HOW THE JEWELRY IS MADE

How crystal jewelry is made and who makes it are two very important considerations. The consciousness of the people who make the jewelry and the place where it is made will be reflected in the jewelry itself. If the people who are making the jewelry are happy, aware of a higher consciousness, and cognizant of the qualities of the stones, you will be able to sense the difference in terms of the vibrancy of the crystals and the piece of jewelry as a whole. Be aware of this as you acquire a piece of jewelry. Jewelry should not only look good, but feel good.

HOW THE STONE IS SET

The way a piece of jewelry is made, and particularly the way the stone is set, is a very important consideration. A stone should be set so that the method of setting does not cut off or impair the energy flow from the stone. For instance, the base or bottom of the crystal generally should not be completely covered. This impedes the energy flow of the stone. This covering is referred to as a cap. Usually the cap is held in place with epoxy glue. This creates a thick sludge-like mass through which the vibration of the crystal must pass. This obstruction impedes the energy flow from the crystal. Practically speaking, if a setting relies entirely on glue to hold the crystal in place, the probabilities are high that you will eventually lose the crystal.

If a stone has been drilled, be sure it has not been drilled horizontally completely through the crystal. This cuts off the flow of energy through the crystal entirely. Drilling or scoring is all right, as long as the stone has not been drilled entirely through in this horizontal direction.

In general, a more open setting is best. This permits the optimum energy flow through the stone and permits natural energy sources, such as sunlight, wind and water, to charge your stone as you wear it. If you continually recharge your stone as you wear it, you will constantly be charging your body with increasing energy. A constantly replenishing circle of energy is created.

METALS

Anything around a crystal influences the stone. Therefore, it is important to select metals that conduct well and channel in useful influences: these are generally silver, gold, copper, or combinations of these. Except in rare instances, don't use anything with lead in it because lead stops or impedes the energy flow of the crystal.

Each metal has characteristics that should be considered and matched with your energy and goals. Silver tends to channel moon, water or feminine energy. Gold channels solar, fire or masculine energy. Copper tends to channel more fire and earth energy. Brass channels primarily fire although it is not as etheric in nature as gold. The vibration of brass is coarser than gold. Metals can be mixed to create a blend of sun/moon or sun/moon/earth energy.

If you are considering a piece of metal that is plated, check to see what the base metal is. Is it a quality with which you want to affect yourself? Does it include lead? Often, an inexpensive plated crystal includes an undesirable base metal.

Steel is sometimes a useful metal to wear with crystals. It connects you with the earth plane and gives you the strength to maintain higher consciousness as you manifest on the physical plane.

Aside from metals, crystals may be set or wrapped in other materials such as leather, fabric, or Native American bead work. These materials will influence the crystal and should be chosen accordingly. Leather influences the crystal both with earth energy and with the qualities of the animal whose skin you are using with the stone. Consider the color and other qualities of the fabric.

How does it feel? Silk, cotton cord, or any natural fiber is appropriate to use. Synthetic materials do not filter and transmit crystal energy well.

How do you choose which metal fits you best? Generally what you are attracted to will be the best for you. Use the same process as you do in choosing a crystal. Notice what seems to attract or draw you towards it. If it attracts you, it is probably resonating in harmony with you. Choose that metal. You may find that you need to change metals as your energies shift. They can shift from day to day, in any other rhythm, or not at all. Consider balancing your body energies. If you feel you have a lot of fire within you and need softening or calming, then choose silver. If you tend to be "airy" or subdued, then select the gold. If you tend to be ungrounded or too etheric, choose copper. Consider what you are trying to accomplish in your work with the crystals. Basically, in choosing a metal, refer to your intuition.

CHOOSING CRYSTAL JEWELRY

You choose a piece of crystal jewelry the same way you would a crystal. Use your intuition and choose what you feel most drawn toward. If you are selecting a piece of crystal jewelry for someone else, concentrate on that person as you study the jewelry. As you concentrate on the person, notice what you are drawn toward. As you focus on them you will start to vibrate harmoniously, and the crystal you choose will vibrate accordingly and be appropriate for that person.

When choosing a piece of crystal jewelry, also notice whether the crystal is polished or left natural. Each has its advantages which should be considered in your selection. A polished crystal tends to create a softer, more rounded field of energy. A crystal which is left in its natural form tends to be more laser-like in its projections. Sometimes, polishing a crystal can bring out its brilliance, adding to its strength. Sometimes cutting and polishing a stone releases more of its power. However, the stone cutter must be aware of the subtle properties of the stone when it is cut. If the cutting does not follow the natural flow of energy the power of the stone will be impeded or cut off entirely. A natural stone is not always better than a cut and polished stone, or vice versa. Each stone is unique and needs to be focused on separately as you make your decision. Finally, as you make your choice, hold the piece for a while and see for yourself how it feels and how it interacts with the various centers of your body. Use your intuition and sensitivity.

CLEARING

Any time you acquire a piece of crystal jewelry, immediately clear it with breath, smoke or salt water before you wear it. (Salt water may tend to oxidize the metal, particularly silver.) An exception can be made if you can sense every influence in the piece of jewelery and want it to remain. Otherwise, clear the crystal of any influence that has entered as a result of the work that has been done on it and the handling of it. For instance, some of the methods used to work with metal include hammering and applying heat. This can have a very jarring effect on you if it is not cleared. Remember, when you clear a piece of jewelry, you are not only clearing the crystal and other stones, but also the metal or any other materials that have been used.

You should also clear your jewelry whenever you have been sick. Also clear it if you have been in an environment that might have transmitted unwanted vibrations into your stones. Clear it if someone touches your jewelry and you don't particularly feel an affinity for them. Usually you should clear antique jewelry.

BALANCE

When you wear crystal jewelry, it is important to be aware of balance. You don't want to throw yourself into a state of imbalance by concentrating too much energy continually into one chakra point. For instance, if you continually wear a stone on your third eye, you can become ungrounded and thus ineffective in the physical universe. So, if you wear jewelry on the third eye or as earrings, it is good to occasionally wear a stone at the navel point for balance. Don't use the amplification capabilities of the crystal to keep yourself in a constant state of high energy without getting rest also. You can tire the physical body and attract illness. If you find this is happening, you can simply take the stones off and not wear the jewelry for a while. Crystals are powerful. When you wear them you have a responsibility to yourself and others around you to assess what is happening.

PROGRAMMING

Crystal jewelry can be programmed just as you would program any crystal. Anything that can be done with a plain crystal can be done with crystal jewelry. The fact that a crystal is set in jewelry makes it convenient to have your crystal with you all the time and to work with it whenever you want. Sometimes cyrstals are sold "pre-programmed." Unless you know and feel an affinity for the person who programmed your crystal it is best to program it yourself. A crystal programmed with you specifically in mind has a stronger effect than one that is more generally programmed.

STORING YOUR CRYSTAL JEWELRY

Use the same careful considerations in the storage of your crystal jewelry as you do with crystals. If you wrap it up to store it, use only natural fibers. You may want to store it on an altar or other special place. If you don't want other people to handle your jewelry, don't call attention to it. Store it in an inconspicuous place. To maintain their highest vibrancy, don't keep your jewelry jumbled up in a messy heap. They are tools. Have respect for them.

COLOR STONES

Crystal jewelry is usually made with clear or amethyst crystals. Not as frequently, the jewelry is made using tourmaline, smoky, citrine, aquamarine, or other natural crystals. To these stones may be fixed other cut and polished colored stones. When a color stone is added to a piece of clear crystal in a jewelry setting, the properties and effects of the color stone are amplified. The resulting effect to you is as if you are wearing a much larger color stone than you actually are.

A clear crystal is the most versatile stone to use. They can be programmed with any color that you would like to work with, and afterward they display the characteristics of that type or color of stone. When you no longer want to work with that color, you can clear it out of the crystal. If you like, you can then program in another color.

Many people like to work with amethyst crystals. They are helpful in any type of healing work. They also influence your body with a spiritual or higher conscious vibration.

When wearing a piece of crystal jewelry near a particular chakra or energy meridian it is often helpful to include with it a smaller stone that has a color which corresponds to that point. This further stimulates these chakra or meridian points. For example, if you are wearing a necklace near the throat center, you might include a turquoise or blue agate with the clear or amethyst crystal. If you tend to have soreness, hoarseness or infection in your throat area, you might wear a more cooling green color that you find in jade, adventurine, or malachite. Rose quartz, malachite or emerald is excellent to add to a clear crystal that will be in a necklace over your heart center. Lapis and sapphire are good to add with any crystal that is affecting the third eye center. Add amethyst to stones on your crown center. Yellow citrines are excellent for the navel area. This stone is very good when used with matters pertaining to will, creativity or physical plane manifestation. Orange stones such as carnelian or madera citrine are excellent for anything which pertains to the second chakra area. Garnets and rubies are excellent for the first chakra energies. Finally, any earth-toned stone is useful for calming and grounding.

It can be useful to refer to guidebooks which show you systems of color stone placement and use. *What is most effective, however, is to be able to sensitize yourself to the effects of color and then apply it to each unique instance.* Again, rely on your own intuition and inner wisdom as you do in all your crystal work.

Finally, the cut of a color stone also makes a difference with regard to its effects. Some stones are stronger when they are faceted because the faceting releases their power. Again, use your intuition and see how a particular stone feels to you. The mark of good jewelry design and work is in recognizing the unique quality of each stone and cutting and setting it in the most appropriate way.

POWER OBJECTS AND CRYSTAL SCULPTURE

A power object is a tool that increases your ability to work with subtle energy and empowers you to do certain tasks that you are called to do. Quartz crystals by themselves are power objects. However, there are ways of combining the quartz crystal with other objects or materials to make a power object that is uniquely yours or is meant for a specific purpose. For instance, many shamanic power objects include specific feathers. Each bird from which the feather comes represents a certain type of energy. An owl is considered a representation of death. To others, that means that it is a bird representing transition. The eagle is the bird that flies closest to the heavens, closest to God and represents that energy. A raven is often regarded as the messenger of the eagle. In some traditions, the raven is like the eagle itself. When a feather is added with a crystal, it adds the quality of that bird to the crystal. Feathers can be used as a wand also.

You also can make crystal sculptures to put in your room. You can have crystals in any environment that you want to affect with the crystal's energy. This is also a way to have crystals in any environment without seeming to be "strange" or in any way inappropriate. Crystal sculpture is a nice way in which the natural beauty of the crystal can be shown to its best advantage.

Custom quartz crystal wand utilizing petrified wood, Macaw feathers, mixed metals and color stones.

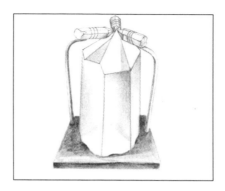

CRYSTAL GENERATOR

This is a particular formation and combination of crystals and metals that create an extremely large and powerful energy field. Use it to flood a large room or even an entire house with crystal energy. Use this anytime that you need the influence of a large amount of energy. The formation of a crystal generator is shown above. Use a large crystal in the center that will stand upright. (At least 6" tall by 4" in diameter.) Place it on a black surface, a holy book, or on any surface which deflects rather than absorbs energy. Choose three smaller crystals of like size. Brace them up so that their tips will point inward to the tip of the large crystal. (The tips can all be touching or the smaller crystals can be slightly apart from the tip of the large crystal.) Place the first smaller crystal where the largest face of the large crystal is. Then place the other two smaller crystals on every other face of the large crystal. Wrap one small crystal with copper, one with silver and one with gold. When you have done this, your generator is ready. Some people like to activate the generator further with visualization, sound or light.

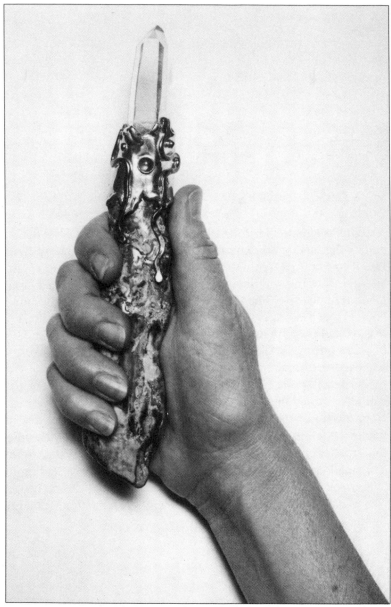

Quartz crystal knife utilizing petrified wood, color stones and mixed metals. Symbolizes cutting through illusion to truth. Also used for psychic surgery and for cutting out and redistributing vibration in the subtle field.

SYMBOLOGY

Certain symbology can be used in conjunction with your crystal. This causes the crystal to vibrate with the attributes of that symbol and similarly affect you. Numbers, designs, pictures, or any other object which has personal meaning to you can be used in a power piece. Power pieces can be worn or carried with you to be used in ceremonial, healing or other appropriate occasions.

The following is a list of some of the more traditional symbols and their meanings. You will often find these included with crystal jewelry.

Triangle -- the trinity -- infinity

Circle -- infinity

Square -- the four directions, finite space, four seasons

Pentagram, or five-pointed star pointing up -- the perfected human being -- the five elements

Solomon's seal or six-pointed star - the union and interaction of matter and spirit or the form and formless

Serpent -- Kundalini energy

Unicorn -- Purity and strength

Sun -- Male principle, light, manifestation

Moon -- Female principle, receptive energy

Crescent moon with the horns up -- The Goddess principle, Isis

Ankh -- The symbol of generation or enduring life, fertility

Heart -- Love

Yin Yang -- Symbol of unity of opposites, the interaction and unity of male and female energy

Om -- Cosmic consciousness or highest realization

Eye of Horus -- Solar male principle, protection against sickness, bringing of life

Eagle -- Union with the highest spirit

Dollar sign -- Prosperity (this can be combined with the Om sign for wisdom to go along with the prosperity)

Trident -- Shiva -- or the realized Being - male creative energy, Neptune

Fish -- Christ spirit

Dolphin -- Cross species communication and joining of consciousness

Cross -- Christ spirit, four directions
Lotus flower -- Purity, realization
Hand with fingers extended, palm forward -- Sign of blessing
Hand with thumb and first finger touching -- gyan mudra - Wisdom and knowledge.

Sun -- Male, solar energy, life force, light of realization of the truth

These are just a few of the more commonly used symbols. Every object can have many meanings attached to it. Each religious group has its own powerful symbology as do many other organizations. Each person has their own private world of symbology. Therefore, the types of symbols are endless. The best thing to do when confronted by obvious symbology is to research the meaning in various reference material, or to see intuitively what it means to you.

SACRED MEDICINE...MIND ALTERING SUBSTANCE AND QUARTZ CRYSTAL WORK

In certain ancient and modern systems of spiritual and metaphysical work the taking of mind-altering substances is practiced as a way to shift your consciousness from a limited reality based on physical sensations of the physical universe to a much more expansive view of reality. When taking these substances, the kundalini energy temporarily rises and the higher energy centers temporarily become energized. When that happens the consciousness and powers associated with those centers temporarily become available to you. You experience an ever-expanding view of the universe. You can learn from and utilize these powers that are temporarily available to you. You can become conscious of and operate on the etheric, astral, mental and other higher planes. You can gain much knowledge from this experience. For this reason some of the shamanic, shiva and other traditions that use mind-altering substances call them sacred medicine. When they are used, it is in a sacred manner. An appropriate environment is chosen, certain Deities, higher guidance or the inner self is invoked with prayer. Certain preparation is undergone and specific practices are followed. Never is this sacred medicine regarded as a "party drug." It is something much more special -- a doorway to the Divine.

As can be seen, there are many positive things that can happen with the use of sacred medicine. However, there are also many drawbacks involved. For many people these outweigh the positive aspects. First, though the use of mild-altering substances can produce a metaphysical experience, it is only a temporary glimpse. You see what is possible, but you do not develop the mechanism to do it for yourself without the substances. Also, the use of mind-altering substances weakens the body. For example, increased amounts of energy can course through your body before you have the strength to handle it. This weakens your nervous system. When your nervous system is weak, you not only tend to get sick but do not have the strength to maintain concentration for any length of time. You often become ungrounded. The psychic muscle system that you need to build up to work on the higher planes becomes weakened or destroyed. Your body energies often become unbalanced. In short, you weaken or destroy just those capabilities which you need to experience higher consciousness without the assistance from these substances.

A substance dependency can be created which further destroys your ability to arrive at these higher states of consciousness yourself. What are you going to do if there is no more of the substance available? (You can't take it with you when you die.) Also, a more subtle dependency often results. You begin to secretly think that you need to take the substances to have access to the higher planes, wisdom, and even your own inner guidance. This not only destroys the confidence you need to do crystal work, but can eventually block your ability to hear your own inner guidance or wisdom.

If you choose to use sacred medicine in your crystal, metaphysical or spiritual work it is best to do it with the following in mind: many practices and exercises that will develop your body, mind and emotions to be able to work with crystals do not mix well with substances. Therefore, it is best to work with a teacher who is experienced with the use of sacred medicine.

Balance is the key word in any work with sacred medicine. Be aware of what is happening with you on all levels. If you find that you are becoming ill, your nervous system is weakening, or you are losing the strength to channel the unleashed energy through your body, stop taking the substances for a while. At the same time, focus even more on doing those practices to regain your health and strength. Build yourself up more than you are weakening yourself. Be aware and be truthful as to any substances' effects.

If you want to use sacred medicine, use it to glimpse what is possible, then use that experience to lead you to work on yourself. Don't use it in place of the work that needs to be done to effectively work with crystals. Build the strength of your nervous system and the rest of your body. Calm your mind and learn how to ground and center yourself. Learn how to raise your energy naturally and be conscious on the higher planes to do your crystal work. It is recommended to use the medicine with an experienced teacher. Above all, use it in a sacred manner.

TAKING CARE OF YOURSELF

When you work with quartz crystals or do any form of metaphysical work, you are working with energy or vibration. You bring into and/or through your body huge amounts of heightened energy in order to do your work. This energy is tremendous. It is the life force of the universe. It is also extremely subtle. In fact, it is easy not to notice the extent of the energy that you pull into your body. This is either because you do not have the development to be aware of it, or because you have become used to it and don't notice it. Because of the strength and intensity of this energy that you use to do crystal work, if your body is not taken care of and strengthened, if you do not regenerate yourself, this energy will ultimately work against you.

As explained earlier, your body should be like a hollow tube through which the energy flows freely, unblocked and unimpeded. You pull the energy in, circulate it in the proper way and send it out, only to pull in more -- a circle of free-flowing energy. If your body isn't strong enough or if there are blockages, this free-flowing "tube" of energy cannot exist. The energy gets trapped or misdirected and tends to lodge in the weakest area of your body system. So, for example, if you are doing healing, the energy you pull from the person that you are healing can become trapped in you, affecting you in various negative ways, or can bounce off you to affect others negatively.

Crystal work, psychic work, healing work, or any work of a metaphysical nature, uses the higher chakras or energy centers: the third eye, crown, throat and/or heart center. When these centers are utilized, the vibratory rate of your body(s) is higher, as is the vibration around you with which you work. To experience this, try the following process: Sit quietly and feel the vibration in the room about you. Don't do anything. Just sit and feel the vibration in the environment about you and in your body. Then meditate on your third eye point. Take long, deep breaths, seeming to breathe in and out of your third eye point. Do this for fifteen minutes. Then, take a deep breath, hold it and let it out. Now, open your eyes and notice what has happened with the vibratory rate around you. Notice the tremendous acceleration of the vibration both around you and your body. This gives you some idea of the energy with which you are working. Many crystal and metaphysical workers are not consciously attuned to the amounts of energy that they are using, to this accelerated vibratory rate. Without attuning yourselves to this and creating a body with the clarity and strength necessary to handle this, you can start "breaking down," physically, mentally, psychically and spiritually.

What are the signs to look for when you're beginning to break down and beginning to be unable to channel this energy? They include the following: You may begin to gain weight. Heaviness, for instance, is often used as armor, or as an attempt to ground yourself. Or the opposite may happen. You may begin to lose weight. Your mind may become unclear or unfocused. You may experience sudden mood swings. You may experience, if you are a woman, disruption or a disturbance in your menstrual flow. A common experience, particularly of a weakened nervous system, is shaking or quaking, either internally or physically manifesting as you work. You may experience body tightness or pain. Your jaw may become tense or locked. Your solar plexus may become very painful or tight. Your neck and shoulders may start becoming painfully tense. You might start having headaches. Your breathing may become more restricted with tightness, irregularity and short, small breaths.

Perhaps you may begin to have changes in your sleep patterns. You may become tremendously sleepy and tired. You feel as if you cannot get enough sleep. Or the opposite may happen. Suddenly you can't sleep at all. Notice if you are starting to rely on stimulants like coffee, black tea or drugs to keep you going, or to keep the realizations flowing through you. At first this seems to help, but eventually it destroys the body. Caffeine and stimulants weaken your nervous system and the entire psychic "muscle system." You may start craving more grounding colors around and on you, or you may start wanting to wear only more relaxing colors like soft green or pink. You may even find that white, blue or purple are too intense. Are you getting messages in your dreams that are telling you that you are getting tired or "burned out"? Are you running into things, pulling muscles or getting hurt? If you are getting into accidents, notice what part of the body you are injuring. What does it seem to tell you? Is a twisted ankle, for example, a message for you to slow down and rest?

What happens when you begin to break down and fail to strengthen or regenerate yourself? First, you start to lose the clarity that is needed for your work. You become inaccurate or ineffective. You become imbalanced. As your body weakens, the energy can not travel up into your higher chakras or energy centers. It starts coming predominantly from the lower centers. If that starts happening, you may start craving power or sexual experience and subtly or blatantly use your work to get it. As your energy lowers, the corresponding consciousness lowers and your lower self gains more control, making excuses for you -- justifying inappropriate behavior. "I'm a realized person. I can do these things. I'm always right because I'm more conscious than other people." There are many variations of this, all believable to your lowered consciousness. Or you may lose your clarity. You become dominated by the ego pictures you have created for yourself. For instance, your ego may say something like, "I'm a great crystal worker. How can I stop doing this. I must continue." Or your ego may say something like "Who am I if I don't work with crystals?" You can forget who you really are and think that you are what you do. You may have created organizations about you and have been trapped by those organizations. They run you instead of you running them. You forget how they were created in the first place. You forget that you can stop, that you can slow down. You fail to notice that you are not pulling in the nourishing energy that you need in order to put out the energy to do the healing or the particular crystal work that you do. The circular motion of pulling in and putting out becomes only putting out, putting out until you arc dry, so to speak.

What can you do if you notice that your body, mind, or spirit are tired? The first thing to do is to *stop*. It takes a certain strength and humility to realize that you are "off track" or stuck, or worn out and then to admit it to yourself and others. Generally speaking, your lower self or ego likes to feel special, that somehow it is above or beyond all of this. ("What will my students think or my friends or clients?" says the ego.)

Realize that nothing that you have accomplished is invalidated, that you are merely mortal, and then stop working. The next thing to do is to strengthen your nervous system. There are several methods to do this, you should try many of them.

First of all, drink lots of water. Drink ginger tea. Take vitamins B and C. You might check to see what other vitamins are depleted in your body. Other good things for the nervous system are bananas, particularly for women. Ginger, cloves and garlic will lend energy to the body among other benefits. Lecithin tends to increase the conductivity of the nerves. To improve the glandular system, you can drink grapefruit juice. A mild diet of beet greens for five to ten days is good for the glandular system, particularly for women. On the eleventh day of the moon your glandular system secretes automatically at its maximum. Therefore, if you fast on the eleventh day of the moon you can strengthen that system. You will probably need to purify the blood. Some things that tend to do that are apples, black pepper, garlic, grapes, oranges, rice and tumeric. (You may want to stay in touch with your doctor during any special diet or fast.)

You should also meditate. Do long, deep breathing. Eleven minutes a day of long, deep breathing through the nose is excellent. It calms and clarifies the mind and charges the body with prana or life force and will relax you on a very deep level. Breathing in and out of the left nostril is very calming. (Similarly, breathing in and out of the right nostril is energizing.)

You might consider a change in your environment. If you live in a city, go to the country. Be among nature, trees and green. Breathe! Sit on the earth. Let it nourish you and heal you.

Consider the colors that you wear. Try wearing a cool, pale green. You might consider wearing earth tones to ground and calm yourself. Wear colors that are very relaxing and cooling. Try not wearing prints and plaids but single colors for a while.

Try doing several exercises that are good for your mind, emotions and body that have been discussed throughout this book. Any exercise will help you.

Also, rest your power objects and store your crystals and stones for a while. Bring them out when you are ready again to use that energy that is generated by them.

Most importantly, notice when you are out of balance, when you have energy bottled up in you, and *go to those sources* that you know can help and *do those things* you know can help and do them immediately. Take this seriously. It is important, not only for you, but for the people that you teach, that you heal and the people who learn from you. You owe it to those you work with to be as clear as possible.

Now, what preventative measures can be used to keep yourself from breaking down and allow you to maintain the proper energy flow? The first thing to do is to ground yourself. It is easy to forget that as you continually activate your higher centers. You need to be a bridge between the planes, to bring that information and higher consciousness down to the earth plane for those you teach or work with. To do that you need to be grounded, to be attached to the earth, so to speak. Also, the earth can transform negative energy to positive and can take any negative energy from you.

Next, develop a support system around you, to support you to do for yourself what you know that you should do. Also, never work against your own rhythm. Work only in tune with your own flow, your own rhythm. Do not let organizations you have created around you or partners you may work with create your method of working or decide for you when you should work. Only work when *you* know you should work. Be in tune with yourself.

Your work takes courage. You must have the courage to be able to stop when necessary, to be able to slow down, to be able to admit that perhaps you do not know everything, to be able to listen to others. You need to be able to see clearly, and not be so attached to your ego projection that you define yourself in terms of what you do. Let go of all of that. Be in the freedom of the moment. Have the strength of steel, the courage of a warrior, and dwell in the joy of an open heart.

You may have known crystal workers, healers or teachers who seemed to have changed or lost something after years of effective work. Though you may have been disappointed or angry, notice how it seemed to happen. Honor them and learn from their experience instead of judging them.

Most of all, it helps to remember the divine play in all of this. It is all illusion on one level. Even the idea of "you" is illusion, and "you doing" is illusion on that same level. As long as you can remember this, you can look at everything with a certain humor and affection. After all, the self is much beyond your body and your work. As you do what you do, you honor and express this self in a way that is unique and wonderful.

ABOUT FOOD

Just as there are many methods and techniques by which you do crystal and other metaphysical work, there are many types of recommended diets. Often, certain techniques require certain diets. If you are doing a particular practice or technique, it is best to follow the particular diet that goes along with it if that diet is an integral part of it.

However, you can develop a sensitivity to food as you develop a sensitivity to subtle energy. Notice how a particular food feels. You can use the same methods that you employ to notice how a crystal "feels" until this awareness becomes automatic to you. Then, notice how your body feels. What kind of food does your body seem to want?

Foods tend to feel heavy or light. They also have the quality of coolness or fire. Meat tends to have a heavy quality, while fruits and vegetables tend to feel lighter. Certain spices tend to lend a fiery quality.

When working to develop sensitivity to crystals and subtle energy, it is good to eat foods that have a lightness to them. Balanced vegetarianism is recommended. Later, as your body's vibration is more fine, you may feel the need for more grounding and meat may then be just the thing for you.

You are not a better or worse person for eating one type of food or another. There is no getting around the fact that you are killing some form of life when you eat. Plants as well as animals have consciousness. You are part of the food chain. However, it is good to thank and honor that which you eat. In a sense, you take on those qualities and consciousness of that which you consume. Their body becomes your body. You might want to offer a prayer that as you take on this animal or plant into your transformed consciousness, that they be in some way positively transformed also. Do what feels best to you.

In summary, be sensitive to the foods you eat. Don't let mere desire, appetite and taste be your only guide in choosing what to eat. First, let your intuitive sense of what is helpful for you be your guide. Then make it taste good!

Do what works for you . . .

4

Crystal Healing

W HAT IS MEANT by the term "healing"? Generally, physical and mental illness, stress and emotional pain and other forms of suffering result from a state of imbalance or disharmony. So true healing is primarily focused on the recreation of natural harmony in and between the physical, mental and emotional bodies. In the process of creating balance, symptoms of the illness disappear. All healing processes that focus on the removal of symptoms also should concentrate on the creation of balance.

There are many methods of doing quartz crystal healing work, probably as many systems as there are people working. As with all crystal work, listen and learn everything. Then do what works for you. Trust and have courage that you are hearing your inner guidance correctly. It is sometimes easy to confuse imagination or intellectuality with intuition. Learn to know the difference between them and have the honesty to admit when you are "off track." Only then can you get back on track and improve your skills. It is sometimes a very fine edge. So constantly check your results. Don't be afraid to admit that you are wrong, or that you made a mistake, or that you could have done something differently. That way, you can keep learning. It is not really you doing the healing anyway, but the spirit that works through you.

Each time you work with a person, you really are laying yourself on the line. You are making yourself vulnerable. These kind of thoughts might start coming up in your mind: "what if this doesn't work: what if they get mad: what if they laugh at me: what if nothing happens after I spent two days doing this healing." It helps to remember that you are offering yourself in doing the healing. You are offering to be that channel. If you can stay out of the way of energy or spirit coming through you, more healing will occur.

ACTIVE HEALING METHOD

The first thing to do in any healing situation is to clear the room, clear yourself and clear your tools. (Refer to chapter on clearing the crystal.) Next, ground yourself and balance your energy. From this state of being grounded, centered and balanced, you will be able to develop an intuitive sense for the vibrations you will be working with. Next, sensitize your hands to physically feel the vibrations of the crystal and the aura or electromagnetic field around the body you are healing. Use the method described on page 21 to do this. Now you are ready to do the healing work with the other person.

When working with another person, first be sure that they are centered. After intuitively checking, if they are not, then center them using the same centering techniques. Similarly, check to see if their energy seems grounded and balanced and guide them through any necessary adjustments.

As you begin the healing, have the person lie down. You might surround their body with amethyst stones and place a rose quartz on their heart center. A single-terminated crystal placed at their feet, pointing outward, will allow unwanted energy to leave their body. Another stone pointing outward from the crown of the head will discourage unwanted vibrations from entering. Do as you feel guided to do.

The next step is to open the person's chakra points and energy meridians on each hand and foot. Use the methods explained earlier to do this. (This process in itself is often enough to balance and heal another.)

Now, regard the person you are working with. What can you sense? Are they open, calm and centered? Do they seem relaxed? Have them relax: have them remember to breathe deeply and slowly.

If you are satisfied, begin the next step: take the crystal in your left hand to energize yourself. Then take another single- or double-terminated crystal in your right hand. Begin to sweep your right hand and crystal through the person's subtle field of energy, about six inches from their body. The crystal in your left hand energizes you, and from that calm centered place where you are working, you sweep through their field or subtle aura from the head down to the feet. Sweep at a speed that best enables you to feel or sense the subtle energy. You should make note of any discrepancies in their field.

If the person is experiencing a specific illness, don't automatically assume that that is the only place you will find discrepancies in their energy field. It may just be a secondary manifestation of another more primary cause. Make note of their complaint or illness but cover the entire body to note any related manifestations.

The kind of discrepancy you may notice includes anything "different" -- a sudden hot spot or cold spot. You may feel a break or dip in their field or areas that seem dense or thick. Make note of these. Be sure to sweep through the entire field, left to right, up and down. You may want to move closer or farther away from the surface of the body. After you have done the front, work on the sides and the back in a similar manner.

If there is something particular that is bothering the person, keep it in mind. You may end up working directly on that area. You may, however, find out that something entirely different is happening. If you don't feel anything you might just try sweeping closer to the body or taking a couple of minutes to re-center yourself. Remember, you are just noting discrepancies in the auric or subtle energy field around the body. Anything that is happening physically with the person is going to show up in the subtle body or their aura. While you are working, continue to retain your relaxed state of concentration. It is sometimes easy to be unfocused and start tightening up. If you notice this happening, take some deep breaths to relax and re-center yourself. Your mind might start wondering onto other thoughts: the most "popular" being those of doubt and questioning: "I wonder if this will work," "Nothing is happening," "I wonder if I"m doing this right"...on and on and on...chatter...chatter...chatter. It is best to just note the thoughts and let them go. Regain your focus. It is natural for your mind to do this. This kind of healing doesn't always make a lot of rational sense and the mind thrives on rational sense. The state you will find yourself working most successfully in is akin to the zen state of "no-mind."

Once you have noted and picked up all of the discrepancies in the subtle body, it is time for the next step. Use a single-terminated crystal (usually clear or sometimes amethyst.) You can use a crystal wand instead because in this technique, the crystal is used in the same manner.

Holding the crystal or wand in your right hand, go back to an area of discrepancy, working systematically from the top to the bottom of the body, or from the strongest discrepancy to the weakest, more subtle ones. Point the crystal or wand at the first place you want to work and start circling clockwise around it. Then feel the crystal drawing your hand around in a spiral to a place right in the center that feels like a little pull or tug. Pull up from that subtle tug as if you were pulling something out of the body with your crystal. Then toss what you pulled out off your crystal and into the earth, where the negativity associated with the illness can be transmuted. Be aware of where you toss -- don't unwittingly toss into your favorite plant, your pet, or transfer the negativity to someone else. You can toss out into the air around you but be very careful to clear the room when you are done in order to transmute any negative energy left in the space. Generally, it is safest to direct the negative energy into the ground.

Work on each area of discrepancy until you sense you have done enough. You will find yourself establishing a rhythm as you do the healing. Sometimes your eyes may close. Remember to stay focused, moving from area to area until you have covered the entire body. This may take an hour, a couple of hours or 15 minutes. Work until you have a clear inner guidance that you have done enough for now.

If, while you are doing this, your crystal starts heating up or seems too dull or in need of some clearing, dip it in salt water to cool it down or clear it.

Or, while using this technique, keep some smoke going from your smudge pot, incense or even a candle. Then you can periodically clear the crystal as you work.

Remember, the key is how focused you can stay. The better your concentration, the better your results.

If you momentarily lose track of what is happening with the body you are working on, sensitize your hands and start again. Using the crystal, you are also energizing the body worked on. Sometimes the auric field may have areas where everything feels slowed down. Sometimes you feel a very speedy vibration. Use your crystal and your mind to even out the vibrational rates of those parts of the body. There will be an intuitive interaction between you, the crystal, and the body you are working on. You will start to feel guided as you work around the body. You'll be drawn to work on a particular area, to go here, go there, do this, do that. As you "hear" or sense these things, listen to them: listen to the inner intuitive voice. The more you work with this voice the stronger it will become.

As you work, pay attention to what your own body is feeling as well. This will also guide you in knowing what the other person is feeling. As you heal, an empathetic connection is developed between you and the other person. As you feel things in your body, you will have even more information about how to proceed.

For instance, you may feel a tightness in your shoulder and realize that the other person is also experiencing that tightness, so you would work with their shoulders. You may feel your body changing in many ways as theirs does. This is another way of channeling the energy. You might feel tingles in your body or a bouyancy. If you start feeling dizzy or unduly tense as you work, take a few long, deep breaths, do a slight or full neck lock to open your upper channels, re-establish your grounding, or drink some water. (Do not drink the water in which you clear the crystal.)

Generally when you are working with someone, it is better not to have any distractions in the room: children, pets, ringing telephones, or other activity. Such interferences can destroy the calmness and focus needed to work effectively.

After you are through, the person you have been working with will be very open. Their natural filtering mechanisms should be activated so they will not indiscriminately pull in all the vibrations with which they come into contact. This is referred to as closing the person off, and can be done in the following way.

Use your breath and blow or use your hand and sweep, horizontally across the body in the area of the heart center. A feather can also be used to do this. Then feel, create, and envision a little wind. You can go from right to left, left to right, or both directions. The person will then be properly set for everyday action outside of the healing space.

Once you have done that, rapidly sweep the body about six inches away from the skin, from the head down to the feet and into the ground. You are sweeping off any negative energy that might be lingering in or on the body into the ground where it can be transmuted. The person you are working on will feel very relaxed. For this technique, you can also use your hands, your breath, crystals, crystal tools or feathers. Remember that the direction to work in is from head to feet. Sweep with vigor.

After you have completed the healing work with the other person, there are some important steps you should do. When you work, the tendency is to pull into your body the negativity that you have pulled out of the body you are working with. You don't want to do that, because if you bring that energy into you it will affect you. This is one of the major problems for healers and makes it very important to learn to circle all the energy out of your body after doing a healing. If you don't rid yourself of the negative energy you have pulled in, you may start getting sick, experience intense mood swings, become nervous or gain or lose weight. You might begin feeling "burned out." You might start feeling tired all the time. Most often, you will get sick.

In order to clear yourself after a healing, begin by washing your hands with cold water. This helps prevent negative energy from moving through the hands into the wrists, and then into your body. Touch the ground again if you like. Then vigorously brush your entire body off while you are standing. Brush from the head down the back, the stomach, to the feet, and into the ground. Touch the ground with your hands. Sometimes, even a moment of connection with the earth is all that is needed. Brush yourself and touch the earth until it intuitively seems to be enough.

Next, clear the room, yourself, and the other person with the smudging or other method as you used when you started the healing. Clear your crystal and any tools you have used.

It is not uncommon for the person you are working on to go to sleep. You can still work with them. Actually, it is not even necessary for the person you are working with to believe that what you are doing is going to help them. All they have to do is be willing to be receptive.

There are some other things to be considered when doing your healing work using this method. At the navel point, there are 72,000 subtle nerves joining, mixing and redistributing energy through the body. Be sure to "tune in" to that area and work with it. That will help subtle energy to be properly distributed through the body. Unexpressed emotions and communications are stored in the body. Unexpressed anger is stored in the navel area. Repressed sadness and repressed communications are stored in the throat area.

If, while you are working, you feel these emotions or communications, then work on the corresponding area of the body. For example, as you are working in an area, you may suddenly feel sadness. Remember whatever feelings you experience because you can later communicate them to the person to help with their healing. These unexpressed feelings and communications often are closely related to what is going on with them.

Also, you should be able to feel energy flowing from the tips of the other person's fingers. If that energy is blocked, you may feel a small ball of energy in their hand and no energy flowing through to the fingertips. The same is true of the bottoms of the feet and the top of the head. They should be open and have energy coming out. We are only balanced when all those centers are open and in harmony with each other. Use your crystals to open these areas and "draw out" subtle energies that should be flowing.

Remember, if your mind chatters, ignore it and continue to stay centered and do your work.

You may work in different ways with regard to verbal feedback. Some people like to have feedback from the person they are working with. They like the other to say things like "I really feel that" or "Nothing much there." Most often, feedback interferes with the work as it breaks your concentration and the other person's concentration. Therefore, you may prefer to work in silence and keep verbal feedback until the end of the healing or until you reach a natural break. Remember, the more focused you can be, the stronger the healing will be.

At the end of the healing, you should not drink or otherwise use the water that was used to cool down or clean the crystals. Pour the water out into the earth. That will transmute the negativity that is left in the water.

When you are healing, you need to be as conscious as possible of everything involved: the work you are doing, the environment around you, your tools, everything.

If you sense your focus drifting while you are working, or if tiredness sets in, take your energizing crystal in your left hand and hold it on your third eye a while. It will help you to focus. Also, while squeezing your energizing crystal in your left hand, call on the forces to help you.

As you are working, observe where people tense up. Are their hands relaxed and open? Is their jaw tense, their forehead scrunched, are they tight across the shoulders? Be aware of these things. You can brush the energy away with your crystal in your right hand. And you will find yourself working differently in each area. While you work in some parts of the body, you will feel very calm and work gently. In other areas, you may find yourself flinging a lot of energy. Feel free to let yourself go even if you find yourself dancing around the person.

Sometimes people heal as you work with them: sometimes it takes several sessions. When you do healing work you should be open to everything. The healing may not look like you expected it to look. What actually may be healing for the person may seem terrible to you in outward appearance. For instance, someone might come to you with a pulled ankle. In working on them, you find the ankle is still in terrible shape, but suddenly the throat seems very open. It may be that the true need in terms of healing was for the throat to open, rather than for the ankle to suddenly be free of pain.

Remember that the spirit going through you does the healing. You are opening yourself to that. Don't invalidate yourself if the healing doesn't look the way you envisioned, or if results aren't immediate. Realize that with practice, things will be easier and results will happen. Also, people often go through illnesses or physical problems for karmic reasons, reasons that fulfill a higher destiny or purpose than is apparent on this earth. If that is the case, they will need to go through what they are experiencing until it has been worked through. Do the best you can: pray for guidance. A useful thought to focus on is "Not my will but thine" and "May I do thy will."

VISUALIZATION METHOD WITH COLOR AND SOUND

The next crystal healing method uses visualization, light and sound. You will be surrounding the person to be healed with large etheric crystals that will provide a protective and energizing field around them. If done with concentration, this is a very powerful technique. It requires the active participation of the person in the healing process. With the spiral method previously described, the person being healed merely needed to be receptive. With the visualization method they are involved in a process which you verbally guide them through.

This healing method will be explained in much detail. It is recommended to first use this method exactly as it is explained for the first few times. Then, if you feel guided to, you can alter it slightly.

Hold a clear, single-terminated crystal in your right hand. In your left hand, hold another clear, smoky or amethyst crystal pointing upward toward your arm. This will provide energy for you. The smoky quartz will help keep you grounded as well as energized. The amethyst will provide healing energy. To start, both you and the person to be healed should sit facing each other. You should both close your eyes, focusing on the third eye and taking long deep breaths through the nose for approximately three minutes. Release all tension from your bodies. Then center yourself and the person to be healed with any of the methods discussed earlier.

Now, with your crystal in the right hand, begin to open the energy centers of the other person. Use methods described earlier. Surround yourself and the other with a field of crystals. This will energize, direct and protect you and the other. No harm can ever come to you in such a space.

Now take the crystal in your right hand and point it into the crown center of the other. Visualize clear light from above going in through that crystal, out the tip and down through the person's crown chakra. Clearly imagine the golden light. This light is going to clear out any blockages, so use your sensitivity. Imagine it pouring down through the top of the head and see if there are any blockages there. During this entire process you can direct the person you are working with as the light comes down, telling them where it is. If they feel any blockage, tension or pain, just have them take a breath and exhale. On the exhale, let that blockage go. Let the light break the density of the blockage, bursting it apart. Now send the light down through the rest of the head, all the way into the neck. Have the other, as well as yourself, experience the head being cool, in clear golden light, with no blockages.

Bring the clear light down into the throat. If you feel any tension as you are working, take a breath, inhale and exhale, letting it go. Pull the light down into the shoulders and down into the chest. Clear the lungs and all the upper internal organs. Feel the light. Feel it down the back, the top of the back, the whole top area. Don't forget the shoulders. All is gold now. Let go of the blocks. Break up any density with light. You might envision grey light moving out of the body as the gold displaces the gray. Continue to direct the other person to do the same visualization process that you are doing. And now, if you feel that all is clear so far, start working down more. With your crystal, guide the gold light down to the stomach and the abdomen. Let the golden light in to loosen and dissolve any blockages. Now move around to the back. Be sure the entire upper, middle and lower back is filled with gold light. Hold fast to your focus. The more powerful your visualization, the better the healing. Pay attention to the arms. Let that golden light first fill up the left arm. Let it come down past the elbow, the forearm, and out the fingers. Imagine all the grey leaving the fingers. Direct it with your crystal if you like. See nothing but streams of golden light out the tips of the fingers. When you've done that, move to the other arm and do the same.

Now go back to the main torso if you've done both arms and go into the lower stomach, the intestines and the navel. Proceed to the top of the buttocks. Past the hips. Remember that with any tension, use your breath. Inhale and exhale, sending the crystal golden light through the blockages. As you are working with the person continually suggest that they do this also. Go down past the sex organs and into the top of the thighs. The entire torso now is filled with golden light. There is no resistance. The light transmutes as it moves through the body. Go into the legs one at a time. Go into the left leg and fill the entire upper part of the leg, the thigh, and the back of the thigh with golden light. And now move past the knee. Remember to see the light in every joint, the skin, the inner muscles, and organs, the blood, everything. Bring the light into the lower part of the leg, into the foot, heel, top of the foot, and center of the foot.

Now imagine the greyness leaving the toes, chased by golden light until you see clear streamers of golden light out the toes. When you've done this, go to the top of the other leg and begin the same process. Direct the golden light past the knee, the bottom of the leg and completely out the foot. Work at your own pace. As the light works directly with your crystal, you may feel like squeezing your crystal. Your crystal may be directing you. You may want a firm grip. You may want a soft gentle grip. The way you handle your crystal may change constantly.

And when you have completely filled the body with golden light, then sit quietly with the person you are working with in this harmonious space.

Now you are going to directly vitalize and balance each chakra point with color and tone. As each chakra is energized, healing tends to take place in the physical, mental and emotional areas associated with it.

Point the crystal in your right hand toward the heart center. You may find yourself actually placing the tip on the heart center or a few inches away. Do whatever seems best. Now visualize green going in through the back of your crystal, out the tip and into the heart. Now sing the tone "AHH" into the crystal, out through the tip and into their heart so that the tone vibrates the green color and fills the heart chakra. Keep doing it until you feel that their heart has had enough for now. You may feel like you want to do it very intensely with laser-like blasts of energy, sound and color. Or you may do it softly, gently massaging the heart with the tone and with the green crystal next move up to the throat area and visualize a turquoise blue color going into the back of the crystal, out the tip into the throat. Lightly rest the tip on the center of the throat chakra. Visualize those turquoise colors and use the tone "OOOO." Vibrate your throat with sound, into the crystal, out through the tip, and into the throat chakra of the person you are working with. Energize it, fill it with turquoise light. When you feel that you've done enough, move the crystal to the third eye point. Imagine the color of lapis, a deep royal blue, streaming into your crystal. Vibrate your third eye with the tone "EEEE," then send it into the crystal and then into the third eye point of the other person. You may find that in working you want to keep your arms straight to channel energy into the crystal. This is much like having a long wand from your shoulder to the crystal. See what it feels like. Other times, you may want to just hold the crystal gently.

Now, move the crystal up to their crown chakra. Point the tip of that crystal into the crown, visualizing amethyst, or purple. Use the seed sound "MMMM" to send the violet color out through the crystal and into the crown chakra. If you have trouble changing colors from center to center, you may want to shake the crystal vigorously, as if you are shaking out the colors. You may want to spend more time envisioning the color leaving the crystal. If you have sage or cedar smoke going as you do this, you can use the smoke to clear the crystal between each different color.

After you feel you've completed this, then intuitively check each chakra. See if they are in balance with each other. If one still feels a little out of balance, correct it.

Next, balance their overall emotional state. For example, if there is sadness, you may want to add a little yellow sun for cheer. You can do this visually or use sound. Imagine the color you are sending through the crystal. What does it sound like? If it had a sound, how do you imagine it would sound? Tone that sound as you continue imagining the color. You may want to channel the energy of a gold citrine crystal into the person.

Next, intuit the mental body or the overall state of mind. How does the person feel mentally? Sometimes you may feel a tight vibration around the head, mental activity and stress, for example. What does their overall mental state of mind feel like? Once you get a sense of it, sense what color it has. And what tone might it have? Then imagine the mental state that would be healing for that person and see the color that state has. What sound does a healed mental state have? Tone that sound in with your crystal through their head or third eye point. You are constantly charging the vibration of the crystal in your right hand with those tones and colors. Those charged vibrations interact with and alter the body's vibrations. This changes the body. In the crystal you have endless tools in one tool.

When you feel you have done enough with the emotional and mental bodies, use the crystal to surround that person completely in clear golden light. The golden aura is circling them, protecting and energizing them. If you like, sense what kind of tone that has. You may sing this tone as you envision the light.

After you have done that, visualize a golden cord of energy from the bottom of their spinal column into the earth. This connects them with the nurturing and transmuting capabilities of the earth. See that they are now a clear channel between the "heavens" through the crown center and the earth through the bottoms of the feet. Finally, as in the healing process before, you want to close the person off and then brush them off from head to toe vigorously. Brush any negative energy, any imbalance, off the body and into the earth. Then wash your hands, and clean yourself, your tools and the space within which you worked.

CHARGING WATER METHOD

Another method that is good not only for healing but also for general health maintenance is to charge water with crystals and other color stones. This technique is also good for energizing your body, giving you a feeling of increased vitality and well-being. To charge water, use a clear glass container and fill it with distilled or spring water. Then choose a clear quartz crystal or any color stone to use with the water. This stone should have a quality or particular vibration associated with it that you feel would be good for your body. For example, if you are feeling ill on a particular day you might select an amethyst crystal. This stone is good for any type of healing that is needed. Similarly, you might choose a clear crystal if you feel the need for more energy. Clear crystals can also be used as a preventative measure against ill health. Before you use the stone, clear it. After you have cleared it you can then program it if you like. Next, take the crystal and drop it into the glass container of water. Sensitize your hands and center your focus upon the container.

Hold your hands, palms down, about three or four inches above the top of the container. If your hands are sensitive and your concentration strong, you will feel a bouyant feeling between your palms and the water. Now, circle your hands, still palms down, clockwise three or four times over the top of the container. With sensitivity you will note an actual change in feeling between your palms and the water with the crystal in it. You now have actually changed the properties of the water to match those of the stone that you dropped in it. You can taste the difference. This includes not only the attributes of the color and type of stone, but any programming you put in it. When you drink this water, it will likewise affect your body in ways inherent to the stone and any programming you might have included. If you are in a hurry you can drink a cup of this water now. However, the effects will be much stronger if you now place the container in the sunlight for an hour, a day or even three days. You can then drink it, or store it in your refrigerator for later use. You might like to prepare in advance several containers of charged water, each with a different color or type of stone that you might later need to use. (If you have programmed any stones, label the different containers so you will know exactly what they contain.) Some recommend putting only one color of stone in each container. Others say that you can mix colors and types of stones. Do what works best for you. When you are healing other people, you can charge water for them that they can take with them and drink over the next few days. Also, be conscious of the effects of the various doses of water. Sometimes a few sips is all that is needed at a time. In acute cases of illness, you might want to drink a glass every 15 minutes for the first hour, then smaller amounts during the rest of the day. Experiment with yourself and check the results.

COLOR HEALING

Begin to learn about color, and when and where to use it. Learn what colors to use with a person in particular circumstances by using your intuitive inner voice while bearing in mind what you have learned in general about color. There are several systems. They will be right for some situations and not right in others. Each healing situation is different. Each person is different.

To develop your own sense of working with color, first consider what is happening with the person you are working with. Then see which color represents or relates to that state. Focusing on the change you want to make, use the color that represents that change. Let's use an example to clarify this. Assume the person has a fever. Close your eyes and center for a bit. What does that fever feel like in your body? What color does it feel like? What comes up in your minds' eye? Choose the first color that spontaneously flashes in your mind. Don't use the rational mind: use your inner guidance. What color comes up? Red: yes, deep fire. Now, imagine that fever gone. What does that feel like? Calm: cool. What color comes to mind? Blue, or sometimes green. Then you take your stone and channel that blue or green light into the area of the fever. Remember, to channel the color, you can use a clear crystal with visualization, or use the actual color stone itself. Sometimes you also surround the person with amethysts for overall healing. Or use clear light to energize. Basically, that is how you work with color. It is just a matter of doing it and doing it and getting feedback on your results. Ask the person a couple of days later how they feel. For more information refer to the section on color and color stones.

HEALING MEDITATION

The following is a crystal visualization method of healing that you can do for yourself or others. Listen to it on tape or have someone read it aloud as you listen to it and follow its guidance.

THE CRYSTAL PATH*

Sit in an upright position. Place a natural quartz crystal in front of you where you can see it. Or, hold the crystal in your left hand gazing into it. If you don't have a quartz crystal with you, you can imagine one. Close your eyes and imagine a crystal. Gaze into the quartz crystal and find a place that seems to interest you and as you gaze go a little closer to that place, look at it closer. What does it look like? Look at it more closely...really concentrate on that place...notice every little detail...pick out one detail and look very closely at it...look...as you look, it seems that you are moving...closer...closer... closer... your eyes close as you move closer and suddenly you are surrounded by the walls of the quartz crystal.

Above you is the tip of the crystal and you sit on its floor. Feel the floor of the crystal underneath you. What does the temperature feel like inside? What does it look

*Transcribed from "Crystal Path" Guided Meditation tape, Uma and Ramana Das, U-Music © 1985   U-103 P.O. Box 31131, San Francisco, California 94131

like? Notice that the light is very clear in the crystal. You see a vibration around you. Very quick, even, protective, vibrating.

Now begin to breathe in and out with long deep breaths. Breathe in filling your lungs, breathe in...and breathe out, emptying your lungs completely. Continue breathing in...out ...in...out...Now as you breathe, notice from the area of your heart, a green light. And as you breathe, notice that each time you inhale the light gets stronger and stronger...You exhale and the light remains stable. Inhale: extend the light: exhale...continue breathing and as you breathe the green light fills the entire area surrounding your heart. Breathe in ...and out...Now notice the area around your throat...as you look it begins to glow with a turquoise light...very brilliantly, the color of sky...turquoise. Breathe in... Breathe out. As you breathe, that color turquoise also grows brighter, brighter, extends further...until the entire area around your throat is bathed in turquoise. Now regard the area between the eyebrows in the center of the forehead. As you look, you see royal blue. Using the breath, breathe in and out ...in and out...in and out...and the area in the center of the forehead glows brighter, brighter, brighter, royal blue. Now notice the area on the top of the head. Looking up you see violet light. You breathe in long, long breaths and breathe out to vibrate this violet light. Extend this light further and further. It shoots straight up from the top of your head as far as you can see. Violet light. Allow the light to spill down around your sides and in front of you and in back of you until you're surrounded with violet light. Completely surrounded. In cool, violet light. Relax in the light. Enjoy its glow. Its feeling. Notice the calmness. The peace. The violet feels very healing. Soft. Gentle. Strong. Violet.

Now notice the edges of your body, skin...that seems to separate you from the violet light outside. As you notice the violet light seems separated from you, start

208 / The Complete Crystal Guidebook

pulling it in to you...its healing...find any area in your physical body that needs healing and pull the violet light into these areas, washing them out, clearing them, no pain, no resistance, gentle ...soft...fill with violet...through every area of your physical body that needs clearing. Then continue to fill your entire body with the violet healing light. As you begin to feel peaceful, content...look at any emotions you may have associated with any illness, any circumstance...and notice those emotions that seem to be particularly painful. Now let the violet light wash through your heart...through your head...through the entire body...soothing you...until you relax. Basking in violet. Purple. Amethyst. Violet light.

Now notice how you feel. Notice how you seem to vibrate in the violet. Vibrations all through you. Now notice the environment around you. How does it feel? Is it harmonious with the violet vibration you have in you? Notice any areas in the environment about you that seem disharmonious, out of balance with you. Fill those areas with violet. Fill the entire environment around you with violet light until you are in harmony. There is no difference between the vibration, the feeling, the wholeness, the peace inside you and the area around you.

Relax in the environment about you. Notice how you feel more content, more peaceful, and joyful. Relax. Now extend the violet around you, outside of you. Extend that violet as far as you can see in front of you to the right side. Extend it behind you further...the left side extend it further...all about you, on all sides...the environment is completely violet. Now extend the violet beneath you. Further, further, as far as you can see beneath you and feel beneath you is violet. Healing amethyst. And above you, look at the violet above your head. Extend it. Vibrate it further up. Higher. Extend it up higher. Until as far as you can see above you is violet. Amethyst. All about you is calm, peaceful. Healed. Completely healed. Bask in the feelings of wholeness. Complete relaxation. Content.

Now, as you look about you and visualize the violet as far as you can see in every direction, start pulling the violet back closer...closer. Notice as you pull the violet closer there still seems to be violet all around you. But you feel like you're pulling back. Closer. Until you notice boundaries about you. The skin of your body. Feel the skin of your body. Notice the pressure beneath you, the floor beneath you. The feeling of the floor beneath you. Feel its validity.

Now notice your breathing. Breathe in...filling your lungs ...hold your breath...and breathe out, emptying the lungs completeley. Breathe in...fill the lungs...breathe out, empty the lungs. Continue breathing, feel your breath. And as you are feeling your breath see the edges of the crystal that you are sitting in. Look around you and see the edges of the quartz crystal in which you sit. Then you notice that the environment around you is all crystal, glowing white. Clear light, high vibration, energizing. Clear quartz crystal. As you sit you look in front of you and a doorway opens, beckoning you out. Visualize walking toward the doorway and you walk out. Now turn and back away from the quartz crystal. The door closes. You watch it close as you back further away. Further away. The crystal seems to shrink and comes toward you. Until you hold it in your left hand. Or remember that it sits in front of you. Now visualize the quartz crystal in front of you or in your left hand. Visualize it. If you're holding the quartz crystal, feel it. Move it around in your hand and see what it feels like. Now open your eyes. If you have a quartz crystal in front of you, see it in front of you...view it. If you have the quartz crystal in your left hand, look at it. Gaze at the crystal. Now notice how you feel. Calm. Peaceful. Content. Healed.

GETTING STARTED AND OTHER
CONSIDERATIONS

Now that some healing methods have been discussed, how do you actually get started? The first thing to do is begin working with yourself with your crystals. Work with your family, pets and even your plants. Start noticing what is around you in your immediate environment. Is there harmony and balance? If so, notice what factors seem to create that. Be honest in your observation. If there is disharmony, what is creating that lack of balance? What can you do with your crystals to re-create balance? If you work first with yourself, your family and your immediate home environment, not only will you learn from actual experience but you will create in yourself and your environment a firm base of basic vitality, health, harmony and wisdom from which to work. Then you will truly have something to offer others. The more you work with yourself, people around you will notice and ask if you can work with them. You will carry a certain presence about you as well as results that can be seen. As you work with others and have results, your healing capabilities and willingness to serve will come to the attention of more people through word of mouth.

You will need courage to do crystal healing work. Even though you may understand that it is the spirit working through you and it will turn out how it was meant to be, others place much different expectations on you. Generally people expect to see certain results when they come to you and probably will not react favorably if they don't see them. Any time you do this work you must open yourself to the healing flow which comes through you and you may feel vulnerable. You might feel as if you are placing yourself at risk. However, when you are centered, you see it is not really a risk. You are still who you are no matter what projections others may place on you. Nothing has really been lost or gained. But it still sometimes feels risky to open up, to be vulnerable, and to do the healing work. You are laying yourself on the line. That is what it feels like to channel. It takes a certain amount of courage to just go ahead and risk it.

It helps when you are working with people to communicate clearly with them before you do any healing work. Explain that you are offering yourself to be a channel and the healing may not look like what they expect. Don't promise what you ultimately cannot control. Offer to try it and see what happens. That helps put it into perspective for others and helps them to be more open to the healing. Sometimes if your approach is one that suggests to others that they obviously can't heal themselves, this disempowers them when actually you want to empower them with the healing. When they are disempowered they are not able to do that which actually makes the healing more powerful: participate with you in the process. So it is good to remember that you are the channel only, and to communicate that to the other person.

Another thing that interferes with crystal healing work and may come up for you is doubt. You have to be willing to leap over it, go ahead and do the healing. There are many manifestations of doubt, so be aware of them when you are doubting and go beyond it. Every healer seems to go through it at one time or another. Some healers go through this every time. Once you start working and the energy starts flowing through you, you become more aware and the doubt disappears. This is similar to stage fright. Just before you go on stage, the fright is the worst. Then, as you are out on stage and performing, it goes away.

Another thing that may come up is fear. It might be fear that you are doing something wrong, fear that you are too "spaced out" in this meditative state, fear that you are not sure what is happening, fear of getting laughed at, or fear of psychic "attack." Realize that if you are putting yourself in the hands of the spirit to use you, you are being guided. You are also being protected. I can't emphasize this strongly enough. This is the case.

There are some things you can do for basic protection if you feel the need to. One is to surround yourself with a golden egg of light and envision that if any negativity comes your way, it will either burst at the edge or be deflected off the edge of this golden light.

The other technique is to surround any object of fear with pink light or sometimes a soft green from your heart. Surround any fear with loving energy. That will take care of it.

Sometimes you can pull in negativity through your heart, sometimes your navel area. If you wear a crystal over these centers it provides a mirror effect. It deflects negativity right back to where it came from. However, the more your heart is open, the less you will have to deal with fear and doubt.

Crystal healing work channels immense amounts of energy through your body and sometimes through the body of the person being healed. A lot of the time, you don't realize it. Sometimes this has certain uncomfortable effects on your physical body which may interfere with your work. For example, sometimes the body will get very cold, either yours or the person's you are working with. In that case, if you are in a room you may want to have it a little warmer. Also, you might want to cover the other person with a blanket and put a sweater on yourself. It is very common for bodies to get cold while working, so have these coverings readily at hand.

Other symptoms that may occur, particularly if your nervous system is weak, are "shakes," tightening of the shoulders, tightening of the jaw, teeth chattering, and developing an unpleasant metallic taste in your mouth. In extreme cases, your whole body may shake. Your nervous system may be too weak for the energy that is coming through. A quick remedy is a loose or sometimes full neck lock. That opens the upper channels and allows the energy to go up and out. To create a neck lock, look forward with your head level. Then lower your chin one inch and pull your head back one inch without dropping it forward. This feels as if the back of your neck is lengthening. Sometimes instead of shaking, you'll get dizzy. The neck lock also works well for dizziness. Then take some long, deep breaths and drink some water. All of this is also a signal that it is time to unblock any closed channels, strengthen the nervous system and possibly rest.

Often, when you start doing your crystal healing work and you start experiencing results, you immediately want to heal the world -- everybody. Do not do that! *Heal only if you are asked.* If you don't wait to be asked, you not only may anger and alienate a person, but drive them further away from any healing. Generally, if someone wants to be healed they will ask-- somebody. You don't have the right to interfere with someone's life. The person probably won't be open to the healing anyway.

If they are not open or they actively oppose you, your healing will usually have no results, which can throw you into a state of confusion and doubt. This will, in turn, adversely affect any future crystal healing work you do for a while.

There is another part to the rule of not healing unless asked. You can heal if or when you have permission. Sometimes you hear very clearly through the inner voice, God, the spirit, whatever you call it, to work with a particular person. Usually if you work with the person without them having asked you and you are working with them because you have permission, don't work overtly as in the first three methods. As you heard in a subtle manner, you work in a subtle manner. You work with visualization. With this method you can heal from a distance. (Of course, you can also heal from a distance if you have been directly asked to do a healing for someone who is far from you.)

In order to do a healing from a distance, first visualize the person. See them clearly in your mind's eye and work on them as if they were with you. On the planes where you are now working, time and distance are of no concern. This type of work takes more concentration than when the person is physically with you. In this type of healing, you are working on more subtle planes. You don't particularly need to understand the workings of those planes, just maintain your visualization and focus.

The person you are healing does not need to know; the person doesn't need to believe that it works or be particularly open to it. So, only heal when you've been asked or when you have permission. Anything else will rebound back to you in a way that makes you wish you hadn't done it. When you try to heal without having permission or being asked, you are operating from your desires, your ego state, rather than in response to the spirits within. This is a difficult rule to follow at times because as you do crystal healing work your heart gets so open, you start seeing all of the suffering, and you ache to help. But it is wisdom to see that sometimes you should not meddle in the way things are. Sometimes that is the best way to serve.

You need to constantly regenerate yourself when doing this work. You can't constantly be "on" all of the time expending energy. You must create a circle of energy which also "refuels" you. To do this you first create the conditions and do those practices which nurture and regenerate you. Then when you are sufficiently "charged" up you can begin your healing work again. By either conforming to a pre-arranged schedule or by noticing the first signs of depletion, you again nurture and re-charge yourself. If you do not, you will seriously deplete yourself and eventually be unable to work. For this reason it is important to *learn to say no*. Don't work when it is not appropriate or not the right time for you to work. Don't work when you are tired. Make an appointment for a later time. Above all, be honest with yourself - - and sensible. Think of a circle of energy when you work, through which energy flows to the person you are working with and back to yourself. You need to take care of yourself. You are not being selfish by declining to work with someone.

It is a mark of a good healer to know when to say no. It is best to be direct and honest with people in declining to work with them, remembering that you still may be able to help. Intuitively sense what is needed to heal the person. Then try to guide them in where they should go or what they might do. You can suggest a simple technique to help them relax and feel better. This does not expend much energy. Toning AHH from the heart is an excellent exercise to show people. It is very calming and will help them see for themselves what they should then do. This empowers them. Many times an illness is a strong message for the person to start listening to themselves. Since all needed information is already within each of us, on some level everyone knows what they need to do. We have only to uncover the needed information. Try to teach people to do things for themselves rather than have them become dependent on you, always having to come back for "a cure." Educate people. Help them become more sensitive to themselves and more responsible in caring for themselves. Be subtle. Don't try to tell them more than they can "hear." Each time they come, teach them a little more.

COMPASSION

What often happens when you do crystal healing work and your heart becomes more open is that you start feeling the suffering in the world around you as well as in those you work with. You may start feeling this physically and emotionally. You may experience it as your own personal suffering, feeling awash with suffering, in anguish, as this level of awareness expands. You may feel like it is too much to bear, that there is no way to help or do enough. Know at this point that there is a reason for everything. There is a goodness in it all, as much as it seems terrible.

You can get another view of suffering by focusing on your third eye point and viewing those situations that are full of suffering or those people who are suffering. View the suffering as you dwell in the third eye point and you will gain wisdom about it. With some of your wisdom you can speak, with some it is best to be silent, but it helps to give you perspective. As you gain in wisdom and do crystal healing work or any work of a metaphysical nature, you will personally experience and understand that everything in the universe fits together perfectly.

The wise ones experience this truth. It does not mean that you should close your eyes or your heart or stop helping. On the contrary. As you develop wisdom, you also develop compassion. Uninvolved compassion. Compassion is not feeling sorry for people. It is acting appropriately and wisely in the face of whatever suffering you see, not as a lesser being or a saviour but as one person to another. Compassion and dispassion are possible when you live in the experience that all life is truly one. Knowing that everything interrelates, you will also know appropriate action as well as proper timing. It is only with the combination of wisdom and heart that you can arrive at this state. Develop compassion by opening your heart and third eye and you will know how to act with effectiveness in true service.

DEATH

Many crystal healers find it hard to deal with situations involving death and dying. For instance, a person may come to you for healing and you may see that they are going to die soon. In fact, you may see that it is right for them to die; it is their time for transition. How do you tell them and how do you work with them? To work with someone in this situation, you have to face your own death. You cannot work with anything you are unwilling to experience yourself. So, before you work with them, meditate on your own death.

DEATH MEDITATION*

Have a quartz crystal standing in front of you at eye level. You should be sitting in a dark room with a light or candle behind the crystal, illuminating it. Sit comfortably with your spine straight. Gaze with relaxed concentration. Continue to gaze into the crystal until your eyes naturally close. As your eyes close, feel as if you are inside your crystal...a protective, safe, enlightening space. As you calmly sit, bring your attention gently to your heart center. Seem to breathe in and out of your heart center. Let any thoughts go as your awareness focuses in your heart. Feel yourself as this calm center and allow yourself to extend through your skin until it feels like there is no difference between the inside and outside of your skin. Extend beyond all your skin: below your feet, out your sides, your front, your back and the top of your head...outward...outward. Feel yourself as having no edges, no boundaries, no body. Dwell in that space...that feeling...that "isness." Can you envision yourself with no body? What does that feel like? What is there? No thoughts, nothing to do. Peaceful floating. What is there? Who is experiencing this? Who are you? As you dwell in this space, slowly become aware of the edges of your skin. Collect yourself through the sides,

*From the cassette tape Flying... Free ©1986 Uma and Ramana Das, U-Music, U-105, P.O. Box 31131, San Francisco, CA 94131

the back, the front, the bottom and the top. Collect yourself until you meet in the center...your heart. Become aware of your breathing and seem to breathe in and out of your heart center. Consider yourself as dying. What emotions come up for you? Sadness...fear? What considerations or thoughts come up for you? Allow yourself to dwell on these emotions and thoughts. Become immersed in them. Now, bring your attention back to your breath. Breathe in and out of your heart center. Consciously and deliberately breathe three deep breaths in, filling your lungs. Each time exhale sharply and completely through your mouth. As you now sit quietly with your eyes closed, visualize the walls of your quartz crystal surrounding you. As you gaze at the walls, envision a door in front of you. In your mind's eye, cross to the door and go out. As you walk further away from the crystal, you see the door disappear and the crystal shrink to its original size. See the crystal standing in front of your eyes. Feel the surface on which you sit. When you feel like it, open your eyes. Sit a few minutes and ground yourself before getting up.

As you meditate more and more on your own death or if you have any conscious dream, astral or other out of body experiences, you learn that being dead is not so bad at all. In fact, it can be wonderful. But the process of dying is what can be difficult. It can be scary and it can be painful. Once you are dead, it is fine. You develop a sense of lightness about it. In fact, being able to communicate that sense of lightness about death while still being aware of the seriousness about it will help a dying person immensely. Be sensitive to the person. Some people are not ready to be told directly that they are dying. If that is the case, in your healing work, work with their symptoms and also work on a deeper level, subtly. Teach them how to be centered, how to be calm minded and less attached to their thoughts and emotions. Help them have experience of their other subtle bodies to loosen their attachment to the physical body. Teach this also to the people to whom you are able to tell directly that they are dying. Work with dreams. All of these techniques are discussed in earlier chapters of this book. As the person goes through their various experiences on the road through transition, be there with them. Be willing to share their experiences, as you would in any effective crystal healing. Be honest with them. Help to make this healing time of growing, self-awareness, opening, and surrender for both you and them. If they are willing, be with them at their time of death. Help them to maintain their state of calm, accepting, self-awareness. Breathe with them. Truly be with them, self to self, through their point of transition. Sit and be with who they are, their self, their soul after they have left their body. Sit in the space with their body, and when appropriate, leave. Send them blessings -- wish them well.

If you like you can chant an ancient mantra for them. It helps send their soul quickly into the higher realms of the subtle planes. You can chant this yourself or have a group of people join you. After chanting this, clear yourself of any attachment the two of you have. This allows them to go speedily on their way. Clear yourself with any method that you would use to clear a crystal. Native American smudging with smoke is good. Clear yourself, the room or space in which they died and any crystals or other tools you have used. If you like, you can charge a special crystal for them that you keep on an altar or outside in nature. If you don't know what to program the crystal with, try one of the following:

> 1. Hold the person firmly in your mind while holding a clear or amethyst quartz crystal. As you do, chant a mantra into the crystal that will help them move to the higher subtle planes and speed them through any lower astral ones. Recommended mantras are OM, RAMA, or SAT NAM. After you chant you can place the crystal on a picture of the mantra's symbol, or wrap the crystal in it.
> 2. Find a picture or draw one of a realized holy being. Focus on the person who died as you hold a clear or amethyst quartz crystal. Ask that holy being if they would be with and guide the departed person through the

after-death realms and help them. If you feel an inner affirmative answer place the crystal on the picture of the holy being. As you do this continually remain focused on the departed person.

After you have done these things, emotionally, mentally and psychically "let go" of the person. Your healing work is done. Sometimes you intuitively hear that you should continue to work with the person after their death. In that case listen to your inner voice. You will hear what to do. You will be guided. You can use your crystals and any techniques that you do with people present in their body with those out of body as long as you clearly visualize them. Listen clearly every day to know when it is time to stop. Then let go.

MANTRA FOR BEINGS IN TRANSITION
To help them on their way
(People, animals, plants)

Stand or sit, eyes closed with your focus on your third eye or on your heart center. Hold your hands in prayer pose and press them firmly against your heart center.

Chant between three and five times slowly (about seven seconds each):

AKAL .... AKAL .... AKAL .... AKAL .... AKAL

The first A in akal is pronounced like "u" in "cup" and is held for half a second or so.

The KAL is held the rest of the time and is pronounced as in "call."

As you do this, you will automatically fall into the correct rhythm.

You only need to do this for one day; however, you can do this for as many days as you like.

You don't go anywhere when you die . . . you are still here now.

5

Further Notes

ATTACHMENT

IN DOING CRYSTAL work, there are inevitably barriers through which you have to pass. As has been mentioned in this book, anytime you cannot concentrate, cannot detach yourself from thoughts and emotions, or cannot hear your inner guidance, you are barred from doing effective crystal work. Another barrier to effective crystal work is when your body is not strong or healthy enough to channel increased energy through it. Methods for surpassing these more obvious barriers appear in this book. There is another major barrier that you must constantly watch for in yourself and overcome many times as it appears in its many guises. It is much harder to detect in yourself because it is much more subtle. What is this barrier? This barrier is attachment.

Attachment is when you focus on false ideas of who you really are and then try to maintain those ideas. Because what you are focused on really is an illusion, and thus impermanent, you will always feel insecure. When you feel insecure you will try to regain your sense of security. Because you are focused on an illusory picture of who you are, (or are not), you will try to maintain this security by false means. You become attached to these false means which, to your mind, seem to be able to guarantee your security. *The result of this form of attachment is always some form of suffering.*

In other words, if you are not focused on who you really are, you will tend to define yourself in terms of what you do, what you think, what you feel, what you have or what you look like. All of these things constantly change no matter how much you try to maintain their permanancy. Your body ages. Your thoughts and emotions may shift. You may be unable to do something that you did before. You are never quite able to be the best at something, there seems to always be someone better. You never seem to have enough, etc. You always seem to feel that there is something lacking or not quite right. Often you don't even consciously acknowledge to yourself that you feel this way. It is just a subtle undercurrent affecting all your thoughts, emotions and actions. You become subject to pride, anger, greed, jealous, and other manifestations of the limited ego.

What is the result of attachment and how does it become a barrier to the most effective crystal work? Fear is one result of this attachment. You fear being wrong. You always have to be right. What if your healing with the crystals doesn't work? When you are feeling fear you become prideful. You feel that you must accomplish more and more and that you have to be the best crystal worker. You become competitive and close your heart. You become too attached to being a crystal worker.

How can you possibly be open and sensititve to others? How can you be open enough to change any method of crystal working to be more appropriate to each unique moment? Eventually you become so concerned with yourself, (your false self), that your primary concern is not really for anyone else that you are working with but secretly to maintain some image of yourself. When you are attached to a false idea of yourself you are unable to completely hear the truth that lies inside you and unable to completely follow the guidance it gives you. Until you can do that, your crystal work will only be partially or not at all effective. To the extent that you remain attached, your higher centers cannot open to be available to you in crystal work. When your higher chakras are closed you are not able to be sensitive to or conscious of the higher subtle planes.

Again, your crystal work will be limited in its effectiveness. When you remain attacted to an illusory idea of who you are you will not be truly fulfilled and deeply content. From this inner fulfillment the best crystal work is done.

It is very easy to get caught in attachment as you do your crystal or other metaphysical work and not know it. This process is extremely subtle. Watch for signs of pride, anger, greed, and other such manifestations. Be honest, don't judge yourself, and be willing to change. To also help alleviate this problem, you can set up a system around you of honest people who will non-judgmentally tell you when they think you are "off track." These people may be close friends or teachers. Look at how you define yourself. Are you defining yourself by what you do? Or by your mind? Are you willing to let go of your old self-definition and find out what is truly there? Going through the various attachments takes courage. Do you have this kind of courage and commitment to yourself and the crystal work that you are doing? The following is a very good meditation to do to work with attachments.

METHOD TO WORK WITH ATTACHMENT

1. Surround yourself with a crystal field: a circle, a double triangle, or any other centering geometric shape.

2. Center yourself and calm your mind.

3. Next, focus on this: Are you willing to be nothing? To not do anything? To be no one special or anything in particular? Just sit.

4. Visualize yourself doing nothing and being of no particular importance. You are not right, nor are you wrong. Just sit.

5. Visualize yourself sitting while life or events just pass you by. Just sit. Let go of having to do anything.

6. Feel what it feels like not to do anything and not be anything. Many emotions and feelings will come up if you do this process with strong intent, concentration and visualization.

How do you feel? Do you feel threatened at all? Do you feel the loss of being something important? What if no one knew you were smart? What if you lived in no place special, doing nothing much? How do you feel? As each emotion, thought or mental picture comes up experience it completely then let it go. Continue the process.

7. After you've done this for at least 11 minutes, take a large quartz crystal and imagine yourself in this state of doing nothing. Enjoy it. Bask in its delight. If you don't feel any enjoyment of this state, pretend that you do.

8. As you imagine yourself in this state of doing nothing and being nothing, notice any thought or feeling that comes up to take you away from the enjoyment of this state. As these thoughts of feelings arise, inhale and exhale blowing them into the crystal with the exhale. Continue doing this. Be aware of any thoughts or emotions that make you feel unharmonious, tense, or upset and blow them into the crystal.

9. Then when you feel like it is time to stop, or you feel complete with this process, take the crystal and clear it or bury it overnight for three days in the earth.

10. Clear the crystals that you had arranged around you and store them. Clear your environment. Ground and clear yourself.

This is an excellent process to go through because what you will learn specifically about are your attachments. Do this process for at least 30 days or any length of time that you choose. Once you know what you're attached to, you can watch it come up again and again during your daily life. Each time, just watch the attachment and see if you can let it go as it comes up. Don't judge yourself. You are not wrong or any less of a person. If judgment comes up, let that go also. As you do this you can also contemplate the question, "WHO AM I?"

Stress is the result of attachment No attachment . . . no stress

CONCLUSION...
THE BEST IS YET TO COME

If you do the exercises in this book until you experience their effect you will develop the ability to be a master crystal worker.

In the process you may well notice other abilities open to you which before might have seemed impossible. You might develop the ability to know exactly what someone is thinking or feeling. You might be able to sense into the past or the future, see subtle colors and auras, hear subtle sound, astral travel and experience various forms of extra-sensory perception. If you use these and other fantastic abilities for your own personal ego advancement, you will lose them and otherwise suffer for it.

What you notice after experiencing these exercises and doing your crystal work is that the crystal and other metaphysical work is not an end in itself. Something else is happening with you that is much more important. You seem to be developing a new consciousness and way of being. You lose the limits within which you placed yourself before. You seem to experience a state of infinite expansion. You begin to live in a state of being which exists without any end or beginning. You are alive in this awareness whether you are awake, asleep, or have left your body. You are empowered and you begin to live in the knowledge of yourself. Anything becomes possible for you as long as it is in harmony with your inner, intutitive voice. Crystal working only increases these experiences. Best yet, you experience an unlimited internal freedom which seem to bubble forth from your heart. You are happy.

You create your own universe

MAY YOU BE HAPPY

May the joy of perfect freedom
permeate the heart
of your soul,

And may radiant tranquility
be yours.

May lightness of spirit
be your gift,

As you follow Truth's vision
with impeccable courage.

Uma Silbey

Instructional and Music Cassettes, Videocassettes,
Natural Crystal Jewelry and Tools from
Uma/U-Music/U-Read

Uma Silbey and her husband Ramana Das have created a number of products and services to augment, expand and deepen the information and experiential processes discussed in The Complete Crystal Guidebook. Their synergistic energies and charismatic relationship has been shared in live performance concerts and sound healing events, lectures and workshops. Currently available products and services are:

Book   Price

The Complete Crystal Guidebook, 240 pages, by Uma Silbey $9.95

Instructional Cassettes

Crystal Path (U-103) 21 minutes each side $9.95
Two guided meditations by Uma, enhanced by ambient synthesized sounds, Tibetan bells and bowls. Side A: Into the crystal's core. Side B: Healing - for yourself and others.

Crystals, Chakras, Color & Sound (U-104) $9.95
Uma and Ramana Das assist you to balance your energies using color visualization and sound practices. The main seven chakras and subtle energy channels are explored, returning to and resting in the heart center. Total time is approximately 40 minutes.

Flying...Free (U-105) approx. 20 minutes each side $9.95
Two guided visualization meditations by Uma enhanced by ambient sounds. Side A: Crystal ball gazing technique into other realms and inner spaces. A flying experience! Side B: Dissolving the barriers of the body/mind. For those in transition, those who are dying, and those who want to practice. Practice makes perfect!

Music Cassettes - Sounds for Voyaging and Healing

Helios (U-101) 17 minutes each side $9.95
Live intense music channeling by Uma and Ramana Das using Chinese gong, percussion, Tibetan bells, bowls and cymbals... for cleansing, healing and re-integration.

Wakan-tanka (U-102) 17 minutes each side $9.95
Ramana Das and Uma, in live performance, take you on a tantric
shamanic voyage into the world of archetypes and inner landscapes.
Orchestral in scope, intense in musical energy, Wakan-tanka
(Great Spirit's mystery) utilizes synthesizers, gong, bells and
bowls, vocal toning, dronal and percussion instruments.

Note: All of the above cassettes are digitally recorded, duplicated in real-
time using highest quality tape. Dolby B Stereo.

Videocassettes
Working with Crystals: visual excerpts from the Complete $39.95
Crystal Guidebook. 50 min. professionally produced vidoecassette of
excerpts from a live workshop led by Uma and Ramana Das
highlighting the major techniques and practices in the book. In one to
three minute segments, these are explained and demonstrated, paralleling
and expanding the written materials in Uma's book. Integrated use of
Tibetan bells, bowls and tincshas bring a musical dimension to crystal
application. Color. VU-7A

Natural Crystal Jewelry and Tools

A four-color catalog of Uma's fine handcrafted crystal and related jewelry
and tools. For your information and direct mail-order convenience.
SASE plus $2.00

Ordering Information

For any or all of the above products, enclose payment by valid personal
check payable to UMA, 655 DuBois St. Suite E, San Rafael, CA.
94901. Include current phone number, driver's license or identification.
California residents add 6% to the total order. Add $1.50 shipping per
item. No extra shipping charge is necessary to order catalog only.
Distributor or wholesale accounts may call 415-453-8845 for
appropriate ordering or information. (Allow 2-3 weeks for delivery
from receipt of order)

Musical concerts, sound healing events, lectures and workshops

The Umananda Institute offers classes, workshops, lectures, concerts
and other events from time to time or on an ongoing basis. For
information write or call: Umananda Institute, P.O. Box 31131, San
Francisco, Ca. 94131 415-381-8865